THE FEELING OF THE FALL

New Directions in Anthropology
GENERAL EDITOR:
Jacqueline Waldren (1937–2021), was Research Associate at the Institute of Social and Cultural Anthropology, Oxford University and Director, Deia Archaeological Museum and Research Centre, Mallorca.

Migration, modernization, technology, tourism, and global communication have had dynamic effects on group identities, social values, and conceptions of space, place, and politics. This series features new and innovative ethnographic studies concerned with these processes of change.

Recent volumes:

Volume 46
The Feeling of the Fall: An Ethnographic Writing Experiment between the Belize Barrier Reef and the Edges of Toronto, Ontario
Ines Taccone

Volume 45
On the Nervous Edge of an Impossible Paradise: Affect, Tourism, Belize
Kenneth Little

Volume 44
Nourishing the Nation: Food as National Identity in Catalonia
Venetia Johannes

Volume 43
Burgundy: A Global Anthropology of Place and Taste
Marion Demossier

Volume 42
A Goddess in Motion: Visual Creativity in the Cult of María Lionza
Roger Canals

Volume 41
Living Before Dying: Imagining and Remembering Home
Janette Davies

Volume 40
Footprints in Paradise: Ecotourism, Local Knowledge, and Nature Therapies in Okinawa
Andrea E. Murray

Volume 39
Honour and Violence: Gender, Power and Law in Southern Pakistan
Nafisa Shah

Volume 38
Tourism and Informal Encounters in Cuba
Valerio Simoni

Volume 37
The Franco-Mauritian Elite: Power and Anxiety in the Face of Change
Tijo Salverda

For a full volume listing, please see the series page on our website:
https://www.berghahnbooks.com/series/new-directions-in-anthropology

The Feeling of the Fall

An Ethnographic Writing Experiment between the Belize Barrier Reef and the Edges of Toronto, Ontario

Ines Taccone

berghahn
NEW YORK • OXFORD
www.berghahnbooks.com

First published in 2023 by
Berghahn Books
www.berghahnbooks.com

Library of Congress Cataloging-in-Publication Data

A C.I.P. cataloging record is available from the Library of Congress
Library of Congress Cataloging in Publication Control Number: 2023007756

British Library Cataloguing in Publication Data

A catalogue record for this book is available from the British Library

ISBN 978-1-80539-034-3 hardback
ISBN 978-1-80539-035-0 ebook

https://doi.org/10.3167/9781805390343

Contents

Figures

Preface

This book is the first of several experiments in ethnographic research and writing about a near-future world beyond catastrophic climate ruin, set in a make-believe, ecotourist community and its offshore coral reefs, evoked by Belize imaginings. This still-too-close-to-the-present, yet "beyond-ruin" climate hovers around the edges of ruin, between the present and an (im)possible future as it echoes between revitalization efforts and prospects, national objectives, group conflicts, individual fears and desires, and experiences of constantly growing and shrinking worlds. By drawing a series of scenes and stories through such an entangled and out-of-time space, the intention of this work is to explore new ways to engage with climate change in Belize and beyond. It is an ongoing project, and as such, this work presents initial and still-developing explorations of experimental writing in the field of climate crisis in anthropology today.

"Beyond ruin" is, in the context of climate crisis today, a felt fluidity, something always moving from the depths of climate ruin. It emerges among things lost or ruined in the course of environmental destruction or deemed too insignificant, inactive, useless, or even too wishful to keep in the face of crisis and emergency, yet still there and leaking beyond any one environmental context in sensations, chats, nostalgic longings, arbitrary memories, dashed plans, small irritations and wonders, sightings, images, and discarded or forgotten objects, trinkets or trash left dangling in history. These are the things that make up the felt losses of climate crisis. When felt as something "lingering" (see Ochoa 2007), and growing with their persistence, however, these things become more than lost. Instead, they move toward curious and lively, as forces reaching toward new relations and potential connections.

"Beyond ruin," thus, becomes excitement and movement in the depths of loss and refusal. It takes bodies and objects someplace else—to another experience, other possibilities, connections, and worlds that draw the historical materiality of ruin into "worlding refrains" (Stewart 2010, also 2007), and potential "global coalescences" (Tsing 2005) that are neither here nor there, yet constantly moving and spilling over. Moving toward what Carla Hustak and Natasha Myers (2012) call "affective ecologies," attending to such movement and liveliness in the depths of loss is to attend to the involvement of difference and various lives in the invention of different ways of living. This work is a modest experiment with such ecologies, as it attempts to reach toward difference in the excesses of ruins in a fictionalized ecotourist, coastal community and its underwater environment, one which moves in and beyond Belize.

This book offers a series of lingering and open-ended fictional scenes and stories that emerge from a process of writing that experiments with unexplored possibilities and impossible connections between worlds and across geographic regions, produced by that which is felt along the edges of ruin as something interesting, uncanny, just plain wrong, silly, or seductive. Such a project is thus interested in future worlds, (im)possible ones potentializing from various experiences with refuse and ruin, that is of loss, across many places and contexts. To take up this task, this book weaves across developmental and environmental issues occurring in two different regions today—that is, I write stories that circle around the touristification of the coral reefs in Belize, while simultaneously being enmeshed in experiences of urban sprawl at the edges of the city of Toronto, and its greater area, Ontario, Canada. The setting, in other words, is thoroughly informed by forces of environmental crisis across regions, as well as images and details of ruin that compound, blend, and diverge into new images and scenes that always remain open. In the process, this book becomes so emmeshed cross-regionally that it is no longer just in Belize or Ontario. It emerges between one place and time and another, while drawing on past experiences of climatic ruin and its capacity to keep going, experiences that are themselves as socially and politically entangled in Belize as they are with global markets and global crises experienced elsewhere. Given this cross-regional approach to ethnographic writing, in which forces of crisis reach across the globe, the following stories take place in a future, in-between place, that is, a make-believe coastal community referred to as "C-Town," in the nation of "Ricuesta," in the year 2040.

By writing with the "beyond" of environmental loss across regions, this work explores and attempts to leave itself open to new thoughts, questions, and connections, which might be made "for better or worse" (Tsing 2005: 267), by engaging with a world beyond what was or is right now—a world in which new entanglements between things wasted, too often forgotten, disposed of, rejected, or deemed too far gone, shocking, or unrelated build onto current, situated climate issues. In so doing, the hope—and it is a hope—

is that the following experimental stories and scenes are different enough from the contexts from which they emerge, while still resonating between regional environmental crises, that they might provoke explorations of potential connections, thoughts, questions, critiques, and problems. As oddly reminiscent, awkward, silly, strange, hairy, problematic, gap-ridden acts of imaginative coalescences and experimental thought, it is hoped that the stories and scenes in this book resonate across places and different groups, environmental groups or otherwise.

The goal of this book has been to explore ways in which experimental, ethnographic work can pick up on moments of "connectability" (Massumi 2002) within climatic despair as "something more" (see Little 2020; Stewart 1996) than just ruin, or what is possible in the present. It is already the case that coastal communities and environmental workers operating off the coast of Belize rely on creative and oftentimes troubling connections and partnerships locally, regionally, and across the planet as they piece their way through the fluctuating deterioration of coral reef systems, as well as the ebbs and flows of tourism. As much as their active world-making works to preserve the reefs and offer prospects of a more sustainable future for sea life, their many entanglements also intensify uncertainty and rupture as environmental actions merge with, shape, and are shaped by the still-changing reefs, development initiatives, community needs and desires, tourist desires, individual desires, intergroup conflicts, and national objectives. As uncertainty and crisis intensify, working "beyond ruin" becomes a continuous effort of picking up the pieces, whether these be environmental, national, animal, material, or social. Yet as a haunting lifeline within growing disaster (Benjamin 1968; Buck-Morss 1991, 1992; Ivy 1995; Little 2020; Ochoa 2007, 2010a, 2010b, 2017; Piot 2010; Stewart 1996; Tsing 2005, 2015), something flows over things ruined as a site of possibility as well as danger, as experiences with them repeatedly spark potential new avenues, connections, and coalescences (Tsing 2005, 2015), new areas of inquiry and concern, as well as desire, which entangle various groups, individuals, projects, and life-forms in new ways. This is already happening. When working "beyond ruin" spans regions, histories, and happenings so different that they can only connect in the imagination, "beyond ruin" may be all the more unsettling, ghostly, and curious as a resonating force with others and difference.

In this book, readers will encounter a series of experimental scenes and descriptions of entangled experiences and speculative connections across two regions, which may emerge as bewildering, curious, confusing, and inconclusive. The work begins from the perspective of an environmental worker in a coastal Caribbean town, one who is also trained as an anthropologist, but diverges into scenes that move beyond any one frame. The scenes have been written with images of things lost, discarded, waning, insignificant, threatened, or threatening due to ongoing environmental issues—yet also reappearing. Although research for this project has also attended to major environmental crises, the focus in the following is on moments and details where the tactility of loss and trouble is most insidious. These moments are the "little things," those small, often mundane experiences, efforts, or things that are cast aside, so common or well known that they become monotonous, but also build in intensity over time (see Cvetkovich 2003; Manning 2016; Tsing 2005). These include moments of longing, ongoing efforts to preserve life amid destructive forces, small disruptions, plans left unfulfilled, passing glances, moments of shock, alliances left corrupted or unrequited, encounters with things forgotten, or too small to remember, and wasted objects, wasted spaces, and even moments where relationships break or threaten to disappear.

By picking up on these moments—moments that are never really lost but always building in intensity within a world of ongoing climatic issues—this work picks up on their liveliness, their capacity to connect, and tries to push what they already are even further by reencountering them in a fictional place. And although writing about such experiences begins with and repeats encounters with loss, the details also reconnect in experimental stories about a place called C-Town. In so doing, the writing becomes an explorative exercise in "Creative contagion," (Massumi 2002: 19), and "speculative pragmatism" (Manning 2016: 2) with climatic ruins in relation to ecotourism in Belize and urban sprawl in Ontario. This work holds close and is inspired by what Kathryn Yusoff (2013) refers to as a "politics of sense," as sensing the "not-yet" of "knotted relations" and worlds as a "transindividual" project (see Manning 2016) that blurs distinctions between here and there, past, present, and future, humans and nonhumans, one experience and another, land and sea, liveliness and wastefulness, loss and production, and development and destruction. By keeping with the many leakages through which "climate" crisis crosses environmental, national, human, animal, material, social, and cultural contexts, the writing for this book blends experiences of loss from one moment to another, to another place or context, on another plane, or in another troubled frame, in scenes that compose creative and experimental entanglements somewhere between a Belizean coastline and the edges of Toronto.

This work thus keeps with forces that whorl up with environmental contexts, as well as the social-political and economic struggles with which environmental issues always entangle. At the same time, writing across regions may bring renewed attention to situated climate issues as the materiality of events and happenings in particular places entwines in fictional ways. In this way, the goal is for the following stories to become something to think about, perhaps laugh about, question, or explore further. Through them I ask: what else might emerge from experiences of crisis, with what is still there, even if degrading, and which time and again constitutes new ways through decline as humans and nonhumans continuously connect, struggle, and take up always changing efforts, actions, desires, or denials in response to climate crisis? Put another way: What new ways can anthropologists write about climate crisis?

More precisely, the capacity to move beyond worlds framed as in crisis or threatened with ruin is explored through creative efforts of engagement with forces and impacts that are felt within them, and which always already set off capacities to move and be moved anew (see Massumi 2002: 15–16). This research attends to such energy, especially where it threatens to break down and yet continually emerges anew in endless struggles faced by humans and nonhumans. The research for this project has focused on encounters in which environmental crisis is prominent, especially notes taken from academic studies, news and magazine articles, films, public documents, or plans that have documented life circumstances and experiences of environmental issues, tourism, and development issues in relation to transformations on land and in the sea in Belize. This research also turns to the context of residential and urban sprawl at the edges of Toronto, and around the Greater Toronto Area, Ontario, Canada. To this, I also add collections of other materials, including notes, journal entries, memories, photos, images, souvenirs, and other objects, which were gathered while traveling through crisis and ruins, wherever these arose, including on tours, nature walks, and city walks, in dumps that are spontaneous or planned, conservation areas, or other public places, and on leisurely drives in Ontario. By encountering these collected materials and images and allowing them to take the writing in this book through interesting and questionable coalescences, experiences of loss are necessarily reencountered, but it is hoped that so too are connections, incipient thought, or just something other within stories about an ecotourist community and its coral reef system, somewhere between the coast of Belize and the outskirts of Toronto.

This book is, therefore, a "speculative" ethnographic experiment about a fictionalized coastal community aligned with a "make-believe" Belize (see Little 2020) and its nearby reef system; and it is an ethnographic form of inquiry into worlds and connections between worlds, not as they exist except in imaginative ways. In other words, the stories do not offer connections already there or waiting to be realized; rather, the hope is to provoke frustrations, questions, or moments of potential connection, no matter how distant or unrelated they may be. As such, all relations and elements that make up the following stories are less organizational components than they are partial encounters with experiences of loss, and in this way, no character, event, group, or even place in the following stories is representative of anything or anyone. Even the endnotes and images are intended to be additions or resonations with the text. The following is, ultimately, a fiction story and an experiment concerned with the beyond and the liveliness of climate crisis and ruin.

Many ethnographers and feminist, post-humanist thinkers have delved into efforts to explore and engage the politics of potential as attending to future-oriented and oftentimes troubling possibilities (see Biehl and Locke 2017; Barad 2003, 2007; Cvetkovich 2003; Gordon 1997; Gregg and Seigworth 2010; Haraway 2008, 2016; Hustak and Myers 2012; Little 2020; Hayward 2010; McLean 2011, 2017; Ochoa 2007, 2010a, 2010b; Pandian 2019; Pandian and McLean 2017; Pine 2019; Piot 2010; Raffles 2010; Stewart 1996, 2007; Tsing 2005, 2015; Yusoff 2013). By taking up a "fictionalizing enterprise" (McLean 2017: 158) in anthropology through storytelling, this work aims to align with these scholars' and ethnographers' efforts by approaching ethnographic research and writing as speculations with the not-yet of environmental crisis and its global reach. As a "mode of active inquiry" (Manning 2016: 10), this work experiments with research and writing as an effort of continuously "adding to" (Massumi 2002: 12–13) emergent reef-tourist worlds from within crisis. As is the case for the contemporary unfolding of environmental disaster, for instance the bleaching of coral reefs, there emerge—within the space of active ruins—seductions left unfinished (Little 2020; Biehl and Locke 2017) that flow across multiple contexts, which time and again bring very different life-forms, and their various needs, desires, and actions, into milieus of new, complex reconstitutions and directions that always move beyond any one reef or shoreline, crisis, or region. Such unfinishedness and possibility emerges in moments that are incomplete and seemingly unreal, but felt. It is hoped that the following, very experimental stories, which echo across regions, contribute in some way to the unfinished possibilities already unfolding within environmental crisis.

Although the research for this project examined studies of social and material entanglements with environmental issues in Belize as they once existed (see, for example, Medina 2010, 2015; Stinson 2014; see also Brondo 2013; Fairhead and Leach 2003 for a small example of related works in the region), the focus of this project is not to help readers sort out such entanglements, nor to analyze the successes and failures faced by certain groups or in specific partnerships. Rather, this book attends to potential connections and entanglements that always (presently, and historically) contribute to the constitution of threatened ecosystems and ecotourist worlds of Belize and beyond it. I thus write in relation to what Gilles Deleuze and Félix Guattari (1987) refer to as "multiplicities" in stories that entwine without capturing experiences of climate crises. By taking this approach, this project becomes a method of research and writing toward what Anand Pandian (2019) calls a "possible anthropology" of "fabulation" within anthropology (McLean 2017), and what Donna Haraway (2016) refers to as "speculative fabulation." At the same time, it includes a continuous exploration toward more within the ruins of climate crisis, without suppressing sensations of them.

As anthropologists, especially those concerned with climate crisis, continue to take up potentializing forces and felt movement as that which always contributes something to the world "for better or worse," this work begins with sensations of loss, and their capacities to break relations, as the very forces of potential to activate new connections and problems (see Tsing 2005). The focus of this work, in other words, turns to the generation and intensification of crisis—the very thing being experienced today as global climate crisis and its capacities to change social and cultural conditions in ways not yet known. It is a short work, and what is hoped for is a starting point to explore ways that climate ruins leak, leach, and return. As an open work, this is an exploration of an inventive process in ways that might contribute to ethnographic writing about climate change—one that is, to repeat, not comprehensive or always sensical, and at times disagreeable, but one that might, however small it is, offer further flows for more thought on experimental writing in anthropology today.

Acknowledgments

I am deeply grateful to all those who have supported, encouraged, and collaborated with me over the course of research and writing. I would like to thank the faculty of the Department of Anthropology at York University for providing the foundations I needed to turn to experimental research and writing, especially Ken Little and Teresa Holmes, who provided years of help and guidance. This project emerged in connection with research conducted for my doctoral studies, supported by a financial contribution provided by the Social Sciences and Humanities Research Council. In addition to my committee members, I would like to thank all those who provided constructive feedback, including the reviewers, who helped to improve the work immensely, as well as Martina Stevenson, who has always been my academic crutch. Kaila Simoneau, who is a brilliant scholar to think with, provided much-needed advice, especially on the images, the former appendix, and photography in general. Thanks also to Dan Savage for his photography skills and support as well. I owe a tremendous amount of gratitude to Sonya, Chris, and James for guidance through the publishing process. And most importantly, to my daughter Sofia, partner Jordan, and family; I am forever grateful for your unending love, support, encouragement, and patience. To the lands, waters, animals, corals, trees, and people who contributed to the following, it is to you that this work is dedicated.

Figure 1.1. "The Fall," north of Toronto, Ontario. © Ines Taccone

Chapter 1

Approximately fourteen hours before the coast of the nation of Ricuesta expected a sudden storm to reach its shores on a day in late September, a humanmade beach was bustling with work and tourist wanderings, at the edge of a small but growing Caribbean coastal community called C-Town. C-Town, with its offshore coral reefs, colorful buildings, and long beaches stretching out from downtown, was an ecotourist town, one quickly expanding on a heightened sense of its own desire . . . even as a storm fast approached. We'd just found out about it the previous day, and at the exact moment I was trying to scrape together another route back to the coast on a renewed work contract with an environmental association, a "global partnership association," or GPA as they were referred to, called CORALREEF. It was the GPA I'd been working with as an anthropologist-turned-environmental worker for the previous six weeks to gather information on locals' and tourists' concerns and desires surrounding CORALREEF's latest revitalization endeavor.

Revitalization of sea life was nothing new in Ricuesta, not with rising temperatures and the intensifying effects of climate change and hurricanes on its fragile ecosystems. The same, however, could not be said about CORALREEF's involvement with what was the most promising and ambitious project ever proposed in C-Town. Everyone was waiting, wondering, and craving to know more about *that* endeavor, especially since it was ramping up faster than expected, as the small nation was in full support. More and more news articles and announcements of public meetings, not just in C-Town but in countries as far as Australia, accumulated throughout the rainy season that year. The news articles always captioned what had become the most popular image associated with the endeavor. I say captioned because the image was really all that mattered. It was a picture of a cruise ship—the very same ship that had been resting oddly close to shore for months—with a large question mark right in the middle of it. Its message was consistently to the effect of: CORALREEF is calling for public input on C-Town's Cruise Ship Project—the effort that would sink a decommissioned cruise ship in the reef system, and around which corals, other sea life, and even tourism were projected to thrive.

But none of the articles explained how this would happen, especially given the shallow waters. All anyone knew, or assumed, was that it would involve that ship, probably in pieces and probably in many places, and numerous new hotels, possibly even in the water. "It's an organic endeavor," a spokesperson said in an interview about the elusiveness of the development plans over the course of the preliminary work already underway. A certain uncertainty was integral to construction, then, with an official announcement scheduled for the end of the season. That, we were informed, was "the best way to proceed."

The project soared before it even got off the ground. The reefs, the size, and the added uncertainty boosted the hype and wonder, especially in C-Town. There was the owner of a restaurant, called Jade, located just one street from the beach; she'd always have one or two tablets with the latest articles loaded, and she'd go around plunking them on random customers' tables with a ridiculously huge smile on her face. It always resulted in her sitting down to chat. She'd do this thing where she'd tap on the screen with her well-manicured fingernails to draw your attention to it, and then mention some very serious fact, which was really *something*, on account of that giant grin. This was someone who named her place after herself, so it was hard to tell if she was actually just excited about a project that would bring more tourists to town, or if she really appreciated having a reason to talk to anyone and everyone. I have to admit, her company was welcomed by customers, who readily swept up those tablets in search of more information, but it was always: "New Partnership with Health Cruises Forged," "Items Removed from Ship in Tubs," or "Possible Impediments to Local Fishing Practices Complicate Managers' Participation." With nothing specific about the actual plans, we couldn't help but get swept away by that image. It made us sweat a little, and it wasn't just the heat.

My contract with CORALREEF was coming to an end, though, and as a result I was trying to figure out my next steps, and not even necessarily in Ricuesta. I'd even had an interview for a new role with a former CORALREEF member, who'd moved on to work with forest regrowth inland in the nearby nation, Belize. The role was only for a few short weeks too, but it was kind of perfect. I was hoping to bridge the gap between my current contract and a possible new one by working on something new too—and anywhere, really. But it had already become clear to me that I'd blown my chances during the interview. Then this storm became . . . something . . . and on the very day—28 September 2040—that I was scheduled to depart C-Town.

The storm took us all by surprise. Mind you, surprises were a fact of life in C-Town, and something to be "rolled with," kind of like how I rolled with that six-week contract with CORALREEF in August that year. CORALREEF taught me much about that kind of rolling. It really meshed with my work, so to speak, and the fact that I never really could describe where I was heading until I was on my way somewhere else, always in passing. Honestly, I couldn't imagine it being any other way. My work is a lot like the way a roller coaster feels . . . after the clicking stops and the cars release into the fall. It's not at all scary; once you get into it, it feels like a rush, and a drop, and it makes you want to laugh. CORALREEF taught me more about that, especially when I stayed in touch with Rita after my initial volunteer work with the GPA two years prior. Rita was one of CORALREEF's managers and a lifelong C-Towner. She was the one who arranged the interview with her former colleague, and this is not even to mention the fact that she'd gotten me that six-week contract. She was always good about staying in touch. That was something to be forever grateful for. I mean, who likes to say goodbye? She truly valued friendships, as was the C-Town way, and, as one of CORALREEF's oldest members, this was something she brought to her "global partnership association" too, including its revitalization of sea life. It had also become increasingly obvious that she was even set on reconnecting with her former colleague and the new forest endeavors. CORALREEF did more than reef work, after all; as was the case for all GPAs, it had once been an NGO, but with all its overlapping revitalization projects, which included partnerships with developers, government, celebrities, royalty—you name it—taking up the "association" turn common among NGOs was the right thing to roll with too.

It also made me wonder how I'd roll with the fact that the interview didn't go as well as expected. I knew how it happened, too—it was when the "smokes," a term used in C-Town for an unidentified odor that haunted the coastal community on and off for years, was brought up. But that's kind of a long story for another time. Suffice it to say that Rita waved it off, saying "It's fine, Beatrix," even though anyone there would have known it wasn't. It was quite a conundrum because Rita was managing the project that I'd been hired to help with, out of which we were to produce a report. With it, Rita was on her way to setting up a new project around fisher families' involvements with CORALREEF and the ship. This, we anticipated, would be fruitful because consultation with C-Town's fisher community was lacking, thereby leaving tensions not only to grow among fisher families but also with CORALREEF and other partners embarking on this project. As part of the report that we'd been working on, I was sure to note that *how* fisher families wanted to be involved was more important than listing concerns and desires. The report would also include an appendix with excess information, notes, queries, parts of interviews, other bits of research, and problems for future exploration, because as Rita said earlier on, "You never know when you might want it." This new project was something I could really roll with, but where it would go, if anywhere at all, was in Rita's hands, and I was hoping to hear back from her about whether she had secured support for it in a month or so.

It also meant I had to bridge the gap. And wouldn't you know it, just as we were trying to produce a potential return to Ricuesta once more, a storm was marked en route to shore. And as was the case on any other day, I did what most tourists dove into once they flooded the beach; that is, we wandered; and so I found myself wandering back onto C-Town's beach, wondering what to do, and what we'd roll with next.

It was my final day in C-Town, and I should've been packing my bags and leaving—running actually—ahead of the storm, especially since I'd already booked a departure ticket before even hearing the news. I probably should've considered myself lucky and taken that flight. Instead, I'd informed Rita that I was too nervous to board a plane with the winds picking up . . . That's what I told her. She probably didn't believe me, because Jade, the owner of that popular tourist-oriented coffeehouse, certainly hadn't.

"It's no laughing matter, Bea," she'd said.

When Rita heard about my plans, she tried to change my mind, saying something about being a foreigner and alone, but finally settled on me staying with her in her apartment. She probably knew that I wasn't getting on that flight, and that it didn't help that we still hadn't gotten the report done—the one that was supposed to be wrapped up by the end of that week. The strange thing was, she didn't even ask about it. She probably knew she didn't need to.

The problem was, I couldn't get my part done—especially with the storm coming. Instead, Rita and I would hunker down, four stories high in one of the more robust condo buildings in town, located on a sideroad off the second main street of C-Town, and ultimately, slightly further from the coast. It was also equipped with storm shutters made of thick rolling slabs of a coarse, fiber-like substance that could not be punctured. More and more newbuilds came with these shutters, including the building in which I was still renting a suite on a flexible agreement. I'd already pulled the shutters down and informed the landlord, who accepted the predicament and merely asked to "lock up" afterward and leave the key with a neighbor. "She knows what to do," he'd said. Still, Rita said she had supplies and that I shouldn't be alone.

She actually said that all the time—to not be alone. Again, she was what others would call a "good friend." It was a little odd that she lived in a tourist condo when locals never did, and she had sisters or siblings who lived together elsewhere. I wanted to ask about that, but it's not something one can ask directly after being gone for two years. What I did know was that Rita was being her usual friendly self, and I appreciated that. That she didn't like the idea of my being alone. It also could be that *she* didn't want to be alone, but who knows. I still appreciated the gesture.

That late September morning in C-Town did not seem too different from any other morning; and in the face of an oncoming storm, C-Towners felt an urge to get to the beach, as did so many tourists. The seafront was only two major intersections over from Rita's suite, opposite lodges, restaurants, and other businesses lining the waterfront. Although synthetic, the beach was far from perfect, yet it remained a gem in the middle of a sprawl of resorts, businesses, and even whole tourist communities sprouting along the waterfront and lengthening the beach in both directions. It bustled with these activities, but nothing more so than tourists who wanted to bask in the beauty of the clear, blue waters whenever they could see them, and the vastness of the blue sky, that advertised feel of the Caribbean. I suppose I wanted to return to the coast to stroll under its tropical trees and watch the waves roll up the shore. I was wondering who was hanging out there; and most of all, it was an opportunity to pick flowers from the plants called "burning bushes," which were planted around the base of trees on the beach. I'd been meaning to pick one of the bright red flowers that bloomed on these bushes and press it in my notebook. Tourists loved to pick these flowers and put them in their hair—which, in fact, was why they'd been planted there in the first place. And then the tourists, after lazily hanging around on the shore, would just as eagerly leave that built beach behind and wander the streets of downtown with no intended purpose—just as I did. And today was no exception.

In fact, it was hard to believe that a storm was on its way, even as it felt too "calm" in C-Town. The sky was gray, but some blue still spotted through the overcast. Stranger still, tourists still hung around the beach, and some were sunbathing. There were telltale signs, of course, like the trees, which rustled in the breeze, and the fact that there were no cruise ships anywhere to be seen on the sea. Not even one. Normally, plateau cruises, the relatively new e-cruises with their ability to come right up to land, could be seen lining the shoreline—except that morning. There was always at least one cruise ship somewhere, even in the distance. At that moment, there was nothing but sky and water. In place of cruise ships, there were white-capped waves—not large—but waves nonetheless, on what were always crystal-clear and calm waters. The water was not as clear either. It gave off a cold, almost stinging sensation on my skin, although the breeze whisking by might have had something to do with it. It was impossible to tell the difference.

Most of the resorts had shut down on the coast, but a few remained open, as some were fully prepared for storms. Practically nothing could stop the higher-end resorts from staying open; and thus, tourists still wandered onto the beachfront. That was what had to be assumed—that the beach tourists were flowing from the still-operating resorts. It would've been no surprise to see at least some tourists, since the storm was unexpected. It wasn't all that surprising, either, to hear talk in town about how more tourists had even flown in from abroad because of the storm . . . That they desired the risk and the experience of a lifetime.

Workers busily covered windows and doors with storm shutters or boards. They set out sandbags, and they carried off lawn chairs, umbrellas, and other items on the beach. They cleared everything but the tourists lying on towels, in chairs, or on hammocks, as if they'd collect the items-in-use later. These sights would have been hard to believe, if not for video footage that later showed up online. The tourists didn't even seem to notice the preparations taking place around them. They just hung onto their hats as they looked out to sea, at the trees swaying in the wind, at the sky, the waves. Some looked at nothing at all as they napped to the sound of waves washing ashore, to the rhythm of a gathering storm.

The hurricane was predicted to reach Category 1 levels. I was not so sure at the time. It could have been less, or more, or maybe nothing at all. No one knew what to think. The only thing to rely on was the disposition to roll with it. And of course, to prepare anyway, when required. We had fourteen hours. And so, wandering ensued.

On the beach that morning, trees busily swayed. A loose coconut tree branch dragged by in the wind, leaving a row of lines in the sand as if trying to hang onto it for a little while longer. This was more effort than that exerted by the flowering bushes around the trees, which seemed to be throwing themselves into the breeze. Since they had bloomed one week earlier and been picked over by tourists, there were already hardly any flowers left. When I arrived, they were almost bare. I remember plucking one of the remaining ones, and the way that its petals blew away one by one, as the thought of what would become of this plant in the storm could not be silenced. And it was there, seemingly from nowhere again, that the smell of something other could be made out in the gusty air. For once, it became apparent that "the smokes" weren't so discernible, yet this smell was. It was the scent of vanilla. I remember looking around but not being able to tell where it was coming from. And there was the surprise of it . . . that it could even be picked up . . . such a delicate scent in the wind.

Suddenly, it came back. I'd been in this position before, only it had slipped my mind. That scent had been there before, right there on that beach, only a few weeks earlier, and that experience repeated yet another experience—a fond experience, and fond memories of encounter—two years prior to that. Now that it comes up, it was like yesterday.[1]

It hadn't been that long since I'd first encountered that exact scent on the beach—just about a month earlier, in August actually, while on an evening stroll. It happened to be a gorgeous, clear night. A waxing silver crescent moon rose over the water, pouring a warmth over the dimmed sands. The clouds had disappeared, and seawater lapped the shoreline. The beach was, for once, unoccupied, except for little crustaceans running sideways in the moonlight with tiny, triangular claws pointing up. A hammock was draped between two coconut trees, where it cradled the silvery sea and sky as if to advertise the forgotten promise of the good life and lull the onlooker back into a dreamy haze. And I remembered someone sitting in the dark, under an umbrella, vaping, not saying anything. Then, there was also the scent of vanilla.[2]

It was difficult to see this person, but their deep, drawn-out inhalations and the sweet smell on that warm night drew the eye toward a cloud of vapor wafting in the dusky sky, where it hovered for a moment and dissipated. It wasn't a creepy feeling; it was just obvious that someone was there. Blending with the humidity, breathing it all in as waves gently washed ashore. It *was* a little disturbing, if you think about it, how it came back on that September morning as the winds picked up and the sea urged forward. But it was an enjoyable beach . . . and it enticed a wish to bask in the beauty that was a memory of it. A shadow of a person was not going to stand in the way of that. But, as I thought about it—and then thought about it again—I'd come across a certain laughing gull on my first trip to C-Town two years prior, and right there on that same beach, that gull threw us all for a loop. On the occasion of our first encounter, there wasn't much to think about. There wasn't much more to think about that morning before the storm either—except to wait and see what would come of the storm, the reefs, the ship, and Rita's project for that matter. Yet still, as if out of nowhere and on the verge of being forgotten, there was a scent and an encounter, and no way to tell how much it could ruin a memory of a warm night on a beach in C-Town . . . that it could take it someplace else.[3]

On the day I first encountered the laughing gull, I was still working as a volunteer with my GPA to educate tourists about specific threats to the reefs. It was 2038, and I was just getting to recognize people and wildlife around the coastal community during my initial visit there—when I got lucky and scored that volunteer gig. I really didn't know much of anything at the time; although an educator, I was very much a visitor. So, when I ran into the gull that day on the beach, my intention was to lump it in with all the rest of the gulls; but on this occasion, the bird darted toward me, staying on my trail as its long, skinny legs scurried across the sand, darting over a couple's legs in an attempt to keep up.

I was so surprised by this animal's unusual character that I excitedly took pictures. As I laughed and cautiously approached the bird, a nearby business owner also approached with a handful of fish scraps. She chuckled as she called him, and called me out on my caution with a frank remark, stating, "You're scared," as she held out her hand for Gregory to eagerly eat from. She leaned toward Gregory with no hesitation, and the bird did the same as he had done so many times before with the hands of those passing by.

Gregory had quite the reputation in C-Town. He appeared in a local magazine as an official resident after Ian, who was known for photographing wildlife, had a photoshoot with the bird on the beach, after picnicking with him almost every day until he moved to another village. It was assumed that this was how he gained the bird's trust, because Gregory did not show up on the beach eagerly or willingly approaching others before these photoshoots. Some said that the bird came to visit Ian after they encountered each other outside of town, a story likely linked to another story told by a local business owner who once saw him feed fish to a laughing gull just steps from her restaurant. She wasn't even sure if it was the same bird, but no matter. The bird started appearing regularly and that's what stuck. No one knew how Gregory got his name either. Some said they heard that "Gregory" was the name given by his old friend, whereas others said he was given the name at random, and eventually associated the name with food. This certainly could've been the case, given that Gregory also responded to the chirping sounds people made at him, a sound he must have associated with something other than his avian mates, since laughing gulls did not sound anything like the humans who called for him. In either case, Gregory's charm was noticed and others in the neighborhood took up the practice of sharing fish with him, myself included.

Gregory stood out in the crowd. This was quite the statement, since all Gregory's fellow gulls stood out with cries that sounded like a cacophony of eerie and annoying laughter. But Gregory wasn't annoying, even if he was, maybe, a little eerie. If C-Town's visitors or community members caught his eye, he'd turn his head one way, then the other, then stop dead center and dart toward them, beak forward, gaining momentum, with his long, skinny legs scurrying over the sand like a body racing on stilts—and then he'd freeze, dead in his tracks about five feet away. Then, he'd wait for them to continue walking, and if they looked back, he darted again, beak forward. This was especially the case if his name was called.

It was eerie, yes. Yet he was also so delightful. No person was immediately receptive to a bird charging at them—who would be?—but somehow, at a certain moment, something different emerged, and it was Gregory who'd be responsible for that difference. Somewhere along the way, this bird established his eternal return with others, before we'd even met. And what was more, he stuck to what became a kind of contest. It was strange, of course. But it gave those of us who encountered Gregory something we were willing to work with. His game was to approach, stop, wait for a response. If all seemed to be going well, that is, if you turned back, he moved closer with a look of curiosity in the way he moved; and if you pulled back, he retreated too. Then, he stopped, waited again, and turned in circles. If walking resumed, he trailed behind. When I stopped, he stopped dead in his tracks, five feet away. When I turned back, so did he. His approach also had a way of rubbing off on the other, and in this way, it also worked the other way around. If he pulled back, I also retreated, and he turned in circles. What a respectful little bird, to mimic the other, and to expend his excitement by running round and round, displaying all angles of himself in the process, instead of forging forward; to flirt and not be so arrogant or insecure as to let that flirtation end at any point. That was when Gregory's forward scuttle moved away from creepy toward "kind of cute," as it brought on our laughter: "Hee-hee!! Look at his little legs go!!"[4]

As unlikely as it was, everyone assumed that Ian taught him the art of attraction. Sorry, I meant to say, they assumed he "tamed" him, but no one could figure out how he did it. All we knew—or figured—was that Gregory's home was likely on the nearby "island," Common Caye, located approximately ten miles southeast of the beach on which Ian likely made his acquaintance.

After meeting Gregory, I started thinking about Common Caye too, especially after having discovered that this caye (the word for "island," pronounced "key") was somewhat mysterious—and was often mistaken for an atoll by tourists. It was no atoll. Common Caye was located not too far from a world-renowned tourist destination located on another caye off the coast of C-Town. Although confused for a natural source of land, Common Caye was not that, nor, despite its name, truly even land. Instead, it was a "dying" coral reef that had got so caught up with infill and dredged materials, rebar, cruise ship waste, plastics, paper products, cell phones, keyboards, old desktop computers, other electronics, construction site waste, household appliances, old mattresses, chairs, sofas, adult diapers, and debris—including entire homes from major hurricanes—that several cayes had surfaced out of the water. Polymers and biodegradable plastics were some of the leading forces of these cayes' emergence. As was discovered later, the partial breakdown of such materials in what are far less than ideal conditions acted like a glue between the refuse caught up in the waters. It became a huge, sedimenting mixture of trash and debris, diapers and plastics, trees and appliances, papers and electronics, and organics oozing across the seafloor. Eventually, these sediments surfaced, and when warmed by the sun, sort of melted together. Then something amazing happened. Mangroves started to grow. And as contaminated materials and garbage accumulated with no other infrastructural development, birds, plants, and other animals—some of which were facing extinction (like the red-footed booby)—made their home in a hurry, in the filth and excesses of urban sprawl.[5]

By the time fishers and scuba divers developed an interest in the area, there was no way to distinguish, at least from under the water, the caye from the shore. Above the water, the caye and shore appeared as two separate places. Underneath, it was clear that the two were connected, only the middle section was (just barely) submerged. The sedimented waste had formed a ridgelike structure, traveling across the seabed and running perpendicular to the shore. The caye, thus, earned the name Common Caye, even as it appeared separated from the mainland above the water's surface. This was, at least, the case at first, because in addition to sea waste, much of the material creating this natural/unnatural phenomenon derived from an extended shoreline produced by the dumping of dredged materials and other waste on a relatively small property about ten miles from C-Town.

For many years, the extended shoreline was used as a secret and not-so-secret dumpsite under the name of a tourist resort development. As more refuse was pushed out to sea, the dying reef also raked it in, and, over time, it all compiled until the cayes merged with the extended length of infilled area off the mainland. As trash accumulated and surfaced, mangroves grew up out of the water, and the Caye, which was no longer a caye, became a kind of tangled mangrove forest clutching the landfill, upon which plants grew, birds nested, and trucks dumped garbage.

The Caye itself wasn't accessible by boat in its early years—with no beach to speak of until one was built years later—but locals soon realized that it was an excellent place to fish, snorkel, and dive around. As it turned out, mangroves were useful for stabilizing the environment, soaking up toxins, and steadying the climate. In fact, the climate around this particular land- and seascape was discovered to be an excellent place for fragile coral nurseries to thrive, and with them, fish and other sea creatures. All sorts of wildlife were attracted there, from essential finfish to the troublesome lionfish. It was at this point that local tour guides saw an opportunity, as did avid scuba divers and fishers. This was an unfamiliar—and what felt like a not-yet-overexploited—site to explore, and the C-Town saying about "Nature's Last Stand" came around.

Back around Gregory went as well, from the Caye to the beach to meet his friends. Tourists fell for the bird on a beach on which they'd then buy lunch, keep the scraps, and feed them to Gregory. Local chocolate shops began selling little white chocolates in the shape of gulls. The more detailed ones were shaded with liquid dark chocolate or vanilla icing. Tourists treasured those chocolates, their lunch scraps, and Gregory. Beachfront restaurant owners were fond of him too, and it wasn't long before Gregory showed up on social media sites as C-Town's wild, beachfront friend.[6]

Notes

1. "A rhizome," write Gilles Deleuze and Félix Guattari, "has no beginning or end; it is always in the middle, between things, interbeing, *intermezzo*. The tree is filiation, but the rhizome is alliance, uniquely alliance. The tree imposes the verb 'to be,' but the fabric of the rhizome is the conjunction, 'and . . . and . . . and . . .'" (1987: 25, emphasis original).
2. I can't help but feel that memories, including this fragrance of a memory, are imitations of a feeling. And that by feeling it, you are passed along and around until you end up right back in the middle of it all, between one place and another, and right here in this book, repeating a feeling recalled on a beach. I think of remembering as a form of imitation, like the anthropologist Michael Taussig wrote: "To get hold of something by means of its likeness" (1993: 21), by mimicking a feeling. And I can't be sure if I held the feeling, or if it holds me and then I get taken away and taken back. In addition to this thought, it might be helpful to note that in what follows, notions of excess and mimesis emerge continually. The works of Michael Taussig, as well as other key anthropologists, and of course, Walter Benjamin, have informed my understandings, which emerge and shift in the following stories. In this way, I attempt to write toward a "systematic openness," as endorsed by Brian Massumi (2002: 17–21), who argued that concepts become productive when they emerge with examples and details.
3. "Things throw themselves together but it's not because of the sameness of elements, or the presence of a convincing totality. It's because a composition encompasses not only what has been actualized but also the possibilities of plenitude and the threat of depletion" (Stewart 2008: 80, following Walter Benjamin).
4. "[T]he eternal return produces becoming-active" (Deleuze 2006: 71). "How is 'becoming with' a practice of becoming worldly?" (Haraway 2008:1).
5. It might be noted that Walter Benjamin's thoughts on history and ruins are caught up with the scenes here. I found Kathleen Stewart's incorporation of Benjaminian concepts and theory in her ethnography *A Space on the Side of the Road* to be especially thought-provoking.
6. Considerable research has been conducted on the impacts of conservation and (eco)tourism on local communities, especially with respect to Indigenous communities, in Belize and surrounding regions, as well as entanglements with the state. Those that stand out most to me are the works of Jill Belsky (1999), Keri Vacanti Brondo (2013), Rosaleen Duffy (2005), Liza Grandia (2007), Teresa Holmes (2010), Jim Igoe and Dan Brockington (2007), James Fairhead and Melissa Leach (2003), Laurie Kroshus Medina (1999, 2003, 2010, 2015), Carel Roessingh and Karin Bras (2003); James Stinson (2014, 2017), and Anne Sutherland (1998). On tourism, nature adventure, and controlled risk, see Kenneth Little (1991, 2020).

Chapter 2

It was just hours before a hurricane was expected to reach C-Town, and the beach bustled along, trees swaying in the wind. Even so, the beachfront was unusually calm given the circumstances. It was hard to believe that a storm was actually coming. Other than the wind, the open and cruise ship-free sea, the hazy overcast sky, and the waves, the day did not feel that ominous; and yet, simultaneously, it did.[1]

Looking back, I should've gotten on that plane, but Rita was finishing the report that would move beyond lists, and quite frankly, how does one leave on that note? Mind you, I could've written it from anywhere. That was the nice thing about CORALREEF. With all their overlapping projects, from the reefs to forestry and even farming, managing an office would've weighed them down, so they didn't have one. It provided the possibility to roll with whatever circumstances came their way, wherever it took them. And yet, there was also something about the beach that took us back to it. And while I faced the vast sea, as vanilla added to the telltale signs of a gathering storm, I was taken back to the night that e-cigarette vapor wafted in the moonlight. It had been two years since I ran into that laughing gull, yet there, in the direct path of an e-cigarette's aromatic gaze, that sense of something forgotten could not be missed. Then, amid another's long, drawn-out inhalation, it rushed back.[2]

After I had returned to C-Town in August on my six-week contract with CORALREEF, I soon recalled my encounter with Gregory two years prior. Curious about his whereabouts, I asked Denise, an acquaintance of mine, whether Gregory still visited the beach. He hadn't shown up for several seasons, she'd said, but I was assured that Gregory didn't just live on the beach. He had to return to his nesting grounds, and those grounds were, it was presumed, on Common Caye, or the extended shoreline. No one knew if that was true. We just wanted to believe it. At least, I did.

I knew Denise as a lifelong C-Town local who'd established herself in a local collective called FTCC. It stood for Friends of The Common Caye, and was a collective that strung together anyone who liked visiting or using the Caye's waters for work or pleasure, and, as a result of the rise of cruise ship tourism, intended to keep development away from the area. The group had partnered with a GPA, the Wilderness Collective, of which FTCC was made a chapter to help manage the Caye with the local community. Denise—through a friend of a friend—connected with a member of the board of the association and landed a job as a coordinator and local stakeholder representative of the Wilderness Collective's FTCC branch. And it was during her time there that she had collected much knowledge about the goings on of the Caye.

"You're sure?" I asked, referring to Gregory's possible home.

Denise chuckled and asked with a raised eyebrow, "Are you?" before assuring me that there was plenty of wildlife on the Caye.

Thus, it wasn't long before I made my way to the Caye in search of a bird I once knew, even if it was highly unlikely that Gregory was still alive; it was still possible, though. That was enough for me. I was no expert on bird longevity, but I did a quick Frankl-e search, and it did say that laughing gulls could live for up to twenty years. It was by luck that I could look that information up, as e-speed internet had just become available in C-Town. When Frankl-e Motors (the once-small electric car company that later extended its focus to all modes of transportation, including internet services) expanded with a vengeance with their revolutionary, wirelessly charged cell phone batteries, the small nation of Ricuesta had to respond with equal vigor. Given that these batteries charged by way of cell tower transmissions, the spread of the new technology among tourists brought with it the need for new towers throughout Ricuesta, and this changed the face of cell phone and internet transmission in the nation just after it had got ahead with the concurrent technology.

It was especially imperative to upgrade and support e-speed internet, as the shoreline faced a rise of cement buildings due to an intensification in the rise of mass tourism. The cement buildings tended to block the signals used to access the internet and charge cell phones by those on the rising number of cruise ships—and later plateau cruises, with their glass floors and e-motors—where tourists anxiously awaited power and connection. So, work had to be done, and C-Town was one of the first places to complete the project. Granted, it required the construction of transmitters, primarily on the water's edge, which, when all was said and done, wasn't the most expensive or difficult endeavor imaginable, especially given the private investment that poured into the pursuit. However, it was still an inconceivably fast move for Ricuesta, and Ricuestians were still celebrating their technological advancement with talk about a "touristic e-state" and "e-thnicity" long before I even arrived. As such, I could gauge whether Gregory was still alive by way of a quick Frankl-e search—although as I later learned, a flawed one, as I came to realize I'd looked up the wrong gull species—and armed with this misinformation, I figured there was an off chance I could find him.

Before my return to C-Town, I hadn't thought to visit Common Caye. Beginning in the 2000s, local fishers and tour guides—who were finding it increasingly difficult to access conservation areas—brought tourists there for a quirky experience, and the number of tourists visiting the site rose thereafter. It was a period filled with uncertainty about the reefs, the state of fisher livelihoods, and the state of the once-successful fishers' cooperative in C-Town as it faced its steady decline. Things were changing. Tourism was increasing. Nothing was predictable. And then, practically out of nowhere, a synthetic island, which had spontaneously surfaced, conjured the desire to explore it. It was still somewhat unknown at that point, and somewhat secret, at least at first.

Although I'd heard of Common Caye at the time of my inquiry into Gregory's whereabouts, I never thought I'd be one to visit because it was mostly a destination for bird and fishing enthusiasts, as well as snorkelers or divers—none of which I was. The section jutting out from the mainland was private property to boot. It mostly attracted expats and tourists crazy or adventurous enough to stumble onto its remains. It was also a big pile of garbage. Let's be clear: it was no surprise that a pile of crap became a desired destination. The expectation for the good life was no more held anymore in C-Town than anywhere else I'd visited. What was left were good life experiences in ruins, sort of like those that the cultural critic Walter Benjamin imagined—but doing something new. During a time when burning "good life" dreams had burned out beyond recognition, there was just as much allure to a toxic, filthy garbage dump—perhaps more—as there was in some promise for the good times that I heard existed somewhere once, but weren't relevant anymore. In C-Town, if promises were made at all, they were promises to get by to the next place of redirection (as when employers alleged: "Don't worry, there will be another contract"), and even those promises were hardly promises, but a statement generally alluding to a stance to "try to try, that is, if all goes well that's what will be done . . . maybe . . . Well . . . we'll cross that road once we get there . . . if we get there." That was necessary, as the common saying was "*Everything is luck.*" I could work with it. We'd learned to roll with it. Dream worlds shifted from those of visions and ideals to "ifs" and "maybes" emerging from what dreams were conventionally defined as: variations of thoughts, images, feelings. All in passing. Potentializing. And so, with that, there were those who liked Common Caye for what it was worth. As for me, as already mentioned, I wasn't *just dying* to visit the Caye. Or at least, I had no reason to go. But I had heard about it, even read about it; and, after asking about Gregory's whereabouts, I had an excuse to actually visit it. What I really wanted, however, was to see Gregory. And the first course of action was to find access to where it pretty much all began, which, as it turned out, was not easy.[3]

The entrance to the infilled dumping grounds, where dredged materials and other refuse were pushed off the coast, was not very accessible at all. The entrance resembled a logging road overgrown by bush, and I wondered how and why anyone would go in there. Yet there it was. I found it. That did not imply going in, however. My next course of action was to approach from the water, by boat, as most others who had come to love the Caye did.

Notes

1. "A story is itself always already a retelling of events, and there is always something more and other to be said" (Stewart 1996: 78).
2. "[T]urn attention, instead, to discontinuity and awkward connection, as this proves key to emergent sources of fear and hope" (Tsing 2005: 11).
3. See Berlant (2011: 24, also 2).

Figure 3.1. Elkhorn Coral, Belize. © iStock, drewsulockcreations

Chapter 3

Common Caye got its earliest start in the 1960s after a major hurricane wreaked much destruction in C-Town and much of the debris gathered by the cleanup ended up being pushed off and extending a small peninsula about ten miles from the town. At the time of the devastation, the property was privately owned by a foreigner who did not even so much as visit C-Town. The owner never opposed the dumping; thus, the garbage stayed there and became a part of the landscape.

Prior to the intensification of tourism in the 1990s, the Caye developed slowly. The area where the debris was discarded was used by many individuals, who dumped garbage there over the years for various reasons, from household waste and cleared brush and trees to home furnishings and improvement debris. By the end of the millennium, it was used for dumping everything from dredged materials to construction site waste from afar.

Decades after the hurricane, the property transferred to another foreign investor named John Tell. John partnered with a development company, with whom he planned to build a beachfront resort with an extended shoreline. He even moved to C-Town. And that was when the infilling really took off.

When I asked colleagues and acquaintances what they knew about the Caye, suspicions surrounding this property came up over and over. For one thing, no one knew why on earth anyone would build a resort out there. It was a dubious undertaking from the beginning. What was more, once construction commenced, it never seemed to progress. It was the slowest-moving development ever. Just when the project was thought to be abandoned another piece would go in. A drive area, the roof, a new wall, a beam would show up. The beam moved. A new beam went in. The new beam disappeared.

"The roof just vanished at one point," Jade said to me one morning over coffee. "So did the walls."

One couldn't help but imagine that the place was going up at the same rate as, if not slower than, pieces of it went missing—that it wasn't just impossibly slow to develop; it was undeveloping.

After several years of ongoing construction, the owners began using the property for other waste materials. Workers came in with trucks and bulldozers to push the already existing refuse heap further into the water, where it would drift and settle as more was piled before it. As the dumping continued, a rough road atop the infilled beachfront was flattened into the emergent landscape, as mangroves grew perfectly along the edges, hiding the interior. Dumping moved out into the water as the pile was pushed further and harder. At first, it was assumed that brush and dirt from the property were discarded off the coast, but it was later learned that the property was being used for dredged materials, mangrove trees, dirt, brush, and demolished homes trucked in from seemingly anywhere. The practice intensified as interest in property development and tourist resorts heightened by the late 1990s and into the 2000s. What no one knew was that, by some miracle, coming up from the water itself, raking in plastics and waste from the sea while mixing with the soils, plants, rubble, and extractions from developments elsewhere, would be a kind of "atoll," which wasn't an atoll but Common Caye, and that it would extend natural-artificial coral nurseries to the shoreline and what would later come to be called Common Senior. Here, tourism and urban sprawl as an extractive industry in Ricuesta repeated itself. In a spontaneous-planned nature–culture partnership, the emergent environment extracted the extractions of tourism development and the cruising industry and reproduced them as another variation of a nature–culture, tourist, and wildlife land- and seascape.[1]

Note

1. It takes watching the removal of forests, hills, wetlands, and rivers to get the sense that urban sprawl, as the anthropologist James Stinson argued about tourism in Belize, is an extractive industry. According to Stinson (2014), ecotourism often results in natural spaces undergoing extraction from their social and ecological milieus, a process that often implies control and destruction. Urban sprawl is akin to such an extractive process, especially when it extracts a space for fantasies of middle-class life by way of tearing up and flattening the earth, and literally paving the way for access by the middle class. Far from producing any sense of ecological responsibility, to use and extend Stinson's argument with respect to ecotourism, urban sprawl "actually produce[s] the opposite" (ibid.: 96).

Chapter 4

When Denise mentioned that Gregory's home was located on Common Caye, I immediately started asking questions about it. That was when I learned more not only about the Caye, but also about Denise. After all, the Caye and the gulls gave Denise her start.

Almost as soon as the contradictory "atoll" emerged from the water, Denise said a flock of thousands of laughing gulls and counting made their way to the newly emerged sea- and landscape, alongside other birds, animals, and sea creatures. And it was because of the birds, specifically the gulls, and later lionfish fishers, that Denise established and reestablished herself with FTCC.

FTCC became a way for Denise to deal with the garbage in her life, as she described it. And she did so while moving beyond that. It was her "shit tool" ex-boyfriend, she said, who put her in touch with a friend who contacted another friend with the ties needed for an interview, which landed her a contract. It was better than, although equally stable as, her previous contracts that she'd cobbled together on resorts, which she hated. She had to renegotiate her position at FTCC on a semiannual basis through what would be the equivalent of endless interviews, but that trouble, she confided, was worth it. Without the gulls, FTCC would have never partnered with the Wilderness Collective, the Ricuestian Conservation Assembly (RCA), and CORAL-REEF, which led to a position for Denise. The Wilderness Collective needed the fishers and tour guides to participate in their newly established fishing and resource use regulations for the Caye, and Denise had lifelong, established connections in the community. What was more, she was also crazy good at interviews, as she explained, since those she had to undergo now were nothing in comparison to interviews she'd prepared for in the past.

So it was that Denise was in touch with goings on around the Caye, and stated that it was assumed that no one went near the garbage heap. That was the case at first, at least. Rumors circulated that John Tell never actually partnered with a developer to build the resort. Instead, they speculated that he was secretly renting out access to other businesses to dump their refuse there while simultaneously working on a construction that seemingly never finished. Others argued that it was another story of a development vision gone wrong and that the garbage had to do with trespassing. In any case, no one knew for sure, and many certainly appreciated the Caye's existence for dumping their own garbage.

It also became known for being a place to hide criminal activities, or government conspiracy. At least, that was what one resident, Jim, believed.

Chapter 5

I was warned not talk to Jim when I started asking questions about him. It was obvious I'd want to talk to him. Those I knew in C-Town called me "a snoop" and they did so for a reason, and now that I think about it, maybe my desire to ask questions also had something to do with the nickname Rita gave me. She called me "Squeaks." I hadn't even mentioned any intention to talk to Jim, but I was still warned not to. They said he was a crazy. They said he would talk my ears off. Then, in case that wasn't enough, they said, "He might be a murderer."

That only made for even more interest. I wanted to know more about this Jim fellow but was too scared to approach him myself. That just added to the wonder even more, followed by even more questions. After talking with anyone willing to chat, I heard one of the strangest stories in C-Town. It was almost as strange—maybe even more—as how Common Caye originated, and it went like this: Jim swore he tripped over the corpse of a sea creature when he went to the property to dump some refuse one day. He said he couldn't identify it, nor figure out how it turned up on land. He said he believed it was a weird-looking mermaid, and probably created in some government lab for the purpose of spying. "Probably the Russians," he'd said.

It was a crazy story cooked up by a crazy-haired expat, who, despite the fact that he was terrified of water, bought waterfront property on which he surrounded himself with local plants in a garden that he constantly tinkered with and redesigned, as his plants were habitually ripped up, hacked down, tossed away, and replaced. He claimed he wasn't afraid of water, just allergic to it. And despite that, Jim would drop his tools, abandon his work, and dart toward the street whenever a wave washed ashore. It was not surprising, then, that a guy who was so afraid of water, yet still chose to live directly in front of it, identified what was likely a misidentified sea creature as a mermaid. As for the spy part—no one knew what Jim was thinking. Everyone shrugged it off, but Jim kept talking about it incessantly. He wanted someone to verify his claim, and he refused to return there to gather evidence himself. He said it was bad luck to see a mermaid twice—a belief he had clearly made up.

His acquaintances didn't buy the excuse. Jim had obviously projected the sea onto the idea of mermaids, and the thought of it scared him into inventing excuses that repressed his phobia further. And because of that repression, Jim talked about mermaids nonstop.

It did not take long for his anxiety and incessant talk about his experience to push two of his neighbors, Jerry and Glenn, to prove Jim wrong. The two went off to see what it was that Jim had actually seen, but then they didn't come back. And when others went looking for Jerry and Glenn, they found nothing, or at least, even less than expected, because the entire resort structure had disappeared! All that was left was a flat space where it once stood. No resort. No sea creature. And no Jerry and Glenn, either. No one could explain it. Of course, this fueled Jim's fears further, and as a result, stories, rumors, and stories of stories circulated thereafter.

That didn't mean C-Towners suddenly believed Jim's claims. Some thought Jim killed his neighbors to keep his delusion going. Others thought something else was going on—that maybe the two trespassers discovered something secret, something that uncovered something, maybe an avenue for smuggling drugs, or maybe something sinister about John Tell. Or maybe, just maybe, there really was something about that sea creature, and that it really was bad luck. From then on, the place was just too strange and suspicious, and most locals would not wander past the entrance thereafter, that is, except for the very adventurous.

Gregory's friend, Ian, was adventurous and ambitious enough to wander into the area for his photos, and everyone knew he personally witnessed the tail end of the Caye's growth. That was because dumping eventually stopped on the property at the same time that John Tell's development project also stopped. Some said that when land-based tourism waned in the area and the rise of cruise ship tourism took over, the property owner simply gave up. Others said he gave up because he was busy with another company that he'd come to represent, which took him inland. Most C-Towners knew, however, that the real driving force behind the end of the business venture and the dumping had much more to do with five major storms.

It starts with a small white spot, and in a few weeks, a beautiful, bright-colored coral that took years to grow can end up as a dead rock with no way back.

—Busiello 2022, on stony coral tissue loss disease.

CHAPTER 6

The day before my contract ended, and before news of the storm reached C-Town, I returned to the popular little coffeehouse called Jade, which I habitually worked out of. Looking back, that day was already gathering the telltale signs of an oncoming storm. It was an overcast and breezy yet still hot morning. It was slightly drier too. News of a storm system to the southeast was being tracked, although there was no real reason for concern at that point. It was also calmer than usual in C-Town—but still busy. Tourists, especially cruise-goers, determinedly wandered across the beach and through the streets as their itineraries were shortened due to the possibility of inclement weather. Many carried flowers they'd plucked from the dwindling burning bushes, while taking snacks and drinks so that they could meet their schedules to get back onboard and get moving.

I was scheduled to depart the following afternoon; in fact, I'd purchased a plane ticket a day or so before. Rita's report still needed to be finalized, though, and it weighed on me. The work had to get to her . . . if I wanted to hear from her later. Yet even so, the most pressing thought on my mind was to get hold of a car I'd rented for the afternoon to explore the outskirts of town. As it was my last full day on the coast, there was pressure to squeeze in as much sightseeing as possible, and exploring the rural areas was still on the list.

When Jade heard about the car rental, she said she wanted to tag along. I'd been working out of her place daily, so we'd become close acquaintances. She even offered her own car, but I wasn't comfortable with the idea. It was a little strange that she even wanted to go, as getting away from work was an unusual desire for Jade, but I didn't ask about it. She hadn't been her usual self since her ex had showed up after disappearing for several months—but that's another story for another time. Perhaps, similarly then, she also wanted to escape the weather, the smokes, the coast. But mostly, I wanted to drive because of the way it made me feel. And I felt like I needed to get away, get over that next hill; in a way, I was relieved about departing.

After all, I had botched the interview that would bridge the gap for my potential return to C-Town, plus I had to get that report to Rita that still wasn't done. The guilt over that was growing too. Rita was known to be very reliable, a good friend who was always there for her colleagues. She was so good about keeping in touch, and there was no doubt that with her new project idea, all I had to do was email the rest of the work to her and wait for a reply.

As Jade and I drove north that day on the main road through C-Town, which was located closest to the coast, we passed between two new neighborhoods being built on either side of the road and bordering the community. Mangroves and other plant life were visibly being removed in portions along the way, providing direct, if spotty, views of what would soon be the beach. Machines were often called in to dredge the seaside to make way for beachfronts for the new residences and lodges. It was conceivable that we would have similar encounters on the way toward the two other developments taking place on the opposite side of town as well. Once all four projects were finished, it would only be a matter of time before more neighborhoods were added. As we drove by the constructions, new structures in a variety of colors were being given finishing touches; we then passed their pale, prefinished, concrete condo forms, which were rising at the edges of the new developments. Beyond these, fields and trees came into view from the open road.

As the rural countryside came into view and the condos passed from sight, I was struck by how quickly we'd driven beyond C-Town. The colors, the transitioning landscape, the speed at which it changed, at which everything was changing . . . it was oddly reminiscent of the North American fall foliage tours that many of the same cruise tourists would just as eagerly seek after their return from their tropical trips. Of course, that was not a beach vacation; rather, fall was a time on the urban outskirts when people, animals, commodities, plants, and leaves rushed across the landscape and crossed paths. It was an undeniably striking moment of passing, those few weeks when the forest canopies turned fiery reds, yellows, golds, and oranges, and sucked their spectators in—and then dropped when they desired. It was a season unto itself, the season between seasons. Fall. A hurricane of fluorescent primary colors swept through the region. And it was during that seasonal transition that the roads—the same ones that connected and gathered always more quarries, houses, and highways—transformed into avenues on which people, animals, trees, leaves, colors, hats, rocks, sticks, rivers, hills, gravel, trucks, cars, garbage, snacks, and coffee mugs crossed and passed each other in the fall. These were avenues for open directions. We became knots in passing on these roads.[1]

The knots in passing were more than that, too. They were magical knots, ones that you could believe would clasp together, but when pulled continuously, they came back out, came undone. To meet on the roads on those fall days—to experience the rush of colors, the warmth of the sun on an otherwise cool day, to pass pedestrians, and animals, and birds flying in formation, and fluttering leaves all moving along the roads, waving "hi" and "goodbye," like they, too, were out to enjoy the fall—was always to meet again on the way somewhere else. They were meetings as magical knots.

Then, we passed by C-Town, where corals continued to return with the help of revitalization efforts, as buildings were erected along the coast, and a cruise ship awaited its new destinations. And then there were the smokes, possible contracts, a possible storm. I felt I needed to get going. I wanted to.

Even as I would've liked to be there when news of the ship was made public, it's not like I could just stay. And it's not as if I'd forgotten that I hadn't had the chance to talk to Rita about her choice of residence—that is, why she lived in a tourist-oriented housing complex when C-Towners avoided those who did. I did know she had siblings in town too, so the self-estrangement was a little surprising. It's just . . . with everything that had to be done within such a short window, it wasn't something easily brought up; and Rita didn't bring it up either, probably for good reason. Not to mention that it didn't sound like it would make a good story—it didn't sound like something that would make anyone want to laugh. And Rita was very good at making people laugh. In fact, Rita was a terrific storyteller. This can't be stressed enough—it was what she left most with people. This was so much the case that it was common to see her at Jade's sitting with a group of people, often tourists, who would seemingly convene as soon as she started talking. It was as if she had a wave in her eye, a don't-be-shy kind of look, or perhaps it was her voice and accent that drew people toward her. What was more, it was as if she could see right into others and reach characteristics of theirs that ran deep but appeared only hazily on the surface; and then, she'd tactfully play with them right in front of everyone. In any other circumstance, that would be unnerving, not to mention that she could be a bit grating with it too, but she somehow rolled it in with laughter and stories.

Some people loved it, but others, like Jade, weren't too fond of it, like how Rita called Jade "Aviator"—referring to a pair of sunglasses she wore years ago and felt quite good in. Then, she'd call her "sweetheart" to boot. She also called me "Squeaks" because one day, when it was so unbearably humid, I sweat so much that my shoes started squeaking against my bare feet. Rita laughed and laughed. I figured that's why she called me that. I found out later that she also thought I complained a lot. It was a really hot month, I swear. It was rare for anyone to ask Rita to stop her teasing, though. It was kind of like adding hot sauce to a burrito. Even if it hurts, you sort of want more; and then you get used to it and can't get enough.

It was no doubt this ability to draw people around her, in addition to her long-term reef-related work, helped Rita secure her manager's position at CORALREEF as well. I can imagine her telling stories and laughing with tourists and environmental workers around the time she established herself in her new role—that is, when a frantic moment involving the reefs broke out in C-Town. It was when a mining company showed up on C-Town's doorstep with a proposal to gut parts of the seafloor, including "dead" corals, with the intention to mash it up to use as white sands for the ever-expanding beaches, as well as a source for aggregates used in building materials. In response, CORALREEF members, including Rita, had to explain again, as they seemingly did every two seconds, that it was better for everyone, including the economy, the nation, the entire world's population, future generations, for the sake of the fish, the sake of common sense, for politicians' sake, for homeowners' sake, for heaven's sake, to keep the seafloor intact.

The aggregate fiasco waxed and waned years before the cruise ship project was even proposed, but it was perhaps in response to that fiasco, or perhaps in response to a lot of little things happening at the same time, that environmental efforts on the coast changed. When various local and international environmental groups rallied against such damage to sea life, they argued that it was more valuable alive—and one of the things that made it particularly valuable was its ability to attract tourists. Once the proposal was pulled, environmental workers had to keep up with and help grow the already expanding tourism industry. Of course, they'd already been doing so, and after the fiasco, the endeavor intensified even more. The cruise ship project figured into that, and Rita, with her expertise and ability to draw tourists to her, made a perfect manager on that team too, even as she, for whatever reason, made an estranged sister that we both avoided talking about.

I wanted to remember the laughter. And then, just after we got beyond the borders of C-Town in a car rented for a last-minute and impromptu rural tour, news of an unexpected, "spontaneous" storm hit the air, and phone calls started pouring in. I suppose Jade and I could've continued on our merry sightseeing way at least for a little while longer; after all, my flight was already scheduled, and what was another, say, hour anyway? Instead, we sped back to downtown—past the developments, the mangroves being removed, a cruise ship awaiting its next course of action, passing the rising buildings transitioning into an array of colorful new abodes, and diving right back into the middle of it all. And with a flight already booked, and a future email waiting for a response from a colleague who, for some reason, lived in a tourist condo, the thought that came to mind was, *So, this is the note I leave on.* And it was obvious at that moment that the storm wasn't the only thing passing through this place. I could imagine Rita saying, "Honey, it isn't even the most pressing."

Notes

1. Kathryn Yusoff (2013) makes important points about knots in her article "Insensible Worlds: Postrelational Ethics, Indeterminacy and the (K)nots of Relating," which echoes in odd ways with the paragraphs here. Her discussion on (non)relation and (k)nots is particularly interesting, especially when she asks: "how might the tying of the (k)not into sense itself be continued to enable responsiveness? And how, if we are to be audacious, might we try to exceed our own sensibility and go on forays into the insensible to gather sense around the insensible spaces between natures?" (ibid.: 217). It should be noted that there is no intention in the stories here to answer these questions.

Chapter 7

The day news of the storm broke out was not the first time C-Towners were caught off guard by unpredictable weather patterns. In fact, the nation of Ricuesta had been experiencing an increase in tropical storm and hurricane activity for decades, and it was only getting worse. So it was that unpredictably, yet not surprisingly, five storms blew back to back into C-Town years prior and ripped the peninsula jutting from John Tell's property out from the mainland. Each storm wasn't horrifically strong, but the accumulated effects of storm after storm were enough to loosen the haphazardly clumped-up waste. The storms ultimately carved a long deep trench at the shore, scattering debris all along the seabed for miles. Separated from its shoreline, the infilled land and mangrove forest once again became a caye, Common Caye. The ridgelike structure reappeared as damaged and shortened, but still there; and what was left jutting from the shore at the entrance became known as Common Senior. Soon after, John Tell severed the section of his property on which he'd planned to build the resort and sold it to another developer.

The Caye, although gaining in popularity, never lost its secretive nature, nor its wonder. This was especially the case because very few wandered around on it, given its inaccessibility and sinister side. For the most part, the first visitors kept their distance, choosing to swim or fish around it. Any visit to the Caye from this distance was still simultaneously an encounter with what was left of a roughed-in road on its way to nowhere in particular, emerging from a combination of the pursuit for profit, tourism, survival needs, and hurricane activity. Most tourists who were taken to this ecological milieu growing through rubble, which became fancied as a "self-revitalizing, secret wilderness," did not know that. They did not know, just as I did not know, that in less than twenty years a mangrove forest grew, nor did they know that Common Caye visitors trespassed if they walked or picnicked, or hid their misdeeds on Common Senior, until they were granted access to Common Caye by the Wilderness Collective in 2035, when a short trail was put in along a portion of the Caye's edge, not far from a small beach-like area that they constructed for touring birders to get onto the Caye by boat. They did not know that the "wild" in the "secret wilderness" had more to do with the wilder and wilder happenings, storms, and developments unfolding in C-Town than it had to do with the fact that the Caye was kept secret for many years. No. The secret had little to do with the past. As much as its past mattered, it had to do much more with the wilder and wilder "-nesses" of the future. Most tourists did not know this any more than they could have known that Gregory would run at them, and that they would take note of the eagerness of his scurry, with his wings slightly lifted and head jutting forward, and that they wouldn't run in the other direction.

Figure 8.1. A view from Tommy Thompson Park, Toronto. © Ines Taccone

Chapter 8

Those engaged in the unofficial use of the Caye, who were also encroaching into the same area that their predecessors once used for fishing and bird-watching, as documented since at least the 1960s, and earlier for fishing and conch-diving, simply fell *in love* with their wilder and wilder "-nesses." The spontaneity was no coincidence, however; "self-revitalizing, secret wilderness" was a phrase that captured the pull their experiences produced; and the animals, specifically the gulls, basked in the love of it.

And so off I went on this wild ride in search of a friend on a caye made up of the excesses of urban sprawl. It was a kind of friendship that differed from other versions of friendship based on lust, or a kind of universal "good." It became more of a story about changing pleasures, changing attitudes.

Friendships, like ones remembered, were events based not on experiences and historical occurrences as they "really were" but on "swirling vibratory relationality," remembered as sensations and bodily patterns in passing, as history, or put another way, "in transition, in variation, when in the act of picking out of the virtual dream worlds [we are] in touch with, [we touch] on a world of feelings, matter and event." In this sense, worlds of friendships emerged from and in a world where distinctions between past, present, and future, here and there, or this body and that were complicated.[1]

Gregory complicated matters. He was a bird I never really got to know, yet he came back to me in remembered worlds that I became in touch with, and which made me want to see him again. When I remembered him, though, he returned not as things really were but as reencountered and still transient passings, melding into stories. Remembering as reencounter did not so much allow me to see Gregory again, then; rather, it made me want to see him again. It reminded me of the "regard" from a distance that I read about in Hannah Arendt's *Human Condition*, that look and consideration that beckoned for his return. Then, he emerged in dream worlds not based in ideals or goals but images and sensations, repeating, collected, and knotted regards in passing reaching for more, and another return.

Some might call this—after Gilles Deleuze and Félix Guattari—a "field of immanence of desire." Perhaps it should be called a future memory. And by remembering Gregory, I let go of the past. I let go of Gregory too as I looked for him again. That desire was not to reconnect with the past as it was, then. Hence, friendship opened into worlds in which remembering enacted sites of new beginning, which for me enacted letting go too, as hard as it was to begin again.[2]

Friendships took us away and took us back, and ironically, their stories had the frustrating tendency to carry many characteristics of a bad old friend. I had many such friends. They weren't always noticeable relationships, since they took years to get old and bad. I recognized them as those I didn't know why I kept around, my crutches made of gelatin—those friends who were fun to bounce off and make fun of, but were completely useless to lean on, and when held up to their own shortcomings, seemed to respond with, "You're welcome."

Notes

1. The quoted phrase and sentence are borrowed phrases—mimicked but not equivalent—from a recent ethnography on a coastal community in Belize. See Little (2020: 138–39).
2. Writing on "the memory of the future," Erin Manning (2016: 50) describes it as "the recursive experience, in the event, of what is on its way. Déjà felt."

CHAPTER 9

The fact that the tourist resort at the entrance to Common Senior never got up and running left the lands above and below the water vulnerable. It was no surprise, then, that a larger development company turned up to buy the property. And this was when a proposal to redevelop a "wasteland" into a site to dock cruise ships, further away from the town but as a way to also expand it, was announced. At the time that this was announced, no one wanted cruises right smack on C-Town's doorstep. Tourists were piling onto the beach. The views were blocked; litter was an issue. Everyone was frustrated. So, this company said, "Don't worry! We'll take them over there, on something that's a miserable wasteland anyway." The plan was to reattach Common Caye to Common Senior, and construct beaches, docking areas, and boardwalks, complete with spaces for vendors for the growing number of cruise-goers to pass on their way to town.

Then something else started.

Up until this point, no one spoke highly of the Caye, if at all. It wasn't terribly popular and certainly not world-renowned. It had no website, no advocates. It was for local use, really. But suddenly, with the news of redevelopment, friends came out of the woodwork, or refuse pit in this case. It just so happened that the fishers who used it (as a source of income through fishing and tour-guiding), as well as expats who fished there, had grown fond of the site. They had noticed a change in fish count; they'd seen the mangroves grow, the birds show up. In reacting to the cruise ship development, these fishers were the first to establish an advocacy group, with their own title and motto, "Fishers for Letting Be Common Caye" (FLBCC). At first, it was only C-Town fishers who made up this group, but as they collaborated with other environmental organizations, advocates, and concerned C-Towners, the group expanded, soon outnumbering the original fishers involved. With increasing numbers in the group from near and far, many agreed it was imperative to acknowledge members' diverse relationships to the Caye with a new, recognizable name. That was when Friends of The Common Caye (FTCC) emerged. It was a friendship born in panic and devotion, as a reactive force—that is, for all of five or so minutes. This was because friendships in C-Town weren't oppositional, they were unpredictable, and, as Denise sort of said about her affairs—love and otherwise—they "mess you up" good too.

Many fishers, feeling rather irked by the new group and its shift in title, retained their connection to the original FLBCC, but agreed to ally with FTCC. They were, in this way, friends of friends of another friend, at least for a little while, all of which was complicated and confusing. And that was the point . . . C-Town friendships were relationships in a constant state of redoing, always beyond themselves, always in need of another rendezvous, as friends say to each other: "Let's catch up." They were an emergent thing, unpredictable and chaotic, freed from determined relations, precisely in that space of the making, the midst of creativity, utterly other at all moments. Relationships were hard. They made us other to ourselves too. They messed us up good, in ways we could only imagine. Then, as if that wasn't enough, friends made us want to see them again.

That was where Gregory came in . . . He came in as an acquaintance before whom something of a history is strung. He was one I still regard as otherwise across space and time, my old friend. He emerged within mine and others' desire as we were taken away and taken back into even more memories that made me want to meet him again. The Caye, as I remember it, was perhaps such a friend too. I can't be certain about the Caye, but I do believe it was more than a theater for friendship. It was the trash and Gregory that I remembered most about the place. And, after all, friends were those remembered—who came along and ruined things. Friendships were complicated, fickle, and somewhat unscripted relationships. Like family members, but not like the structure of family at all, they really got on one's nerves.

Not to mention into them. No one wanted to swallow water around the Caye. This was not because of the salt content, but because despite the Caye's amazing ability to hold itself together, it did ooze toxins into the water, and some of those toxins were known to cause central nervous system damage. What kind of friend does that, you ask? Possibly only the best.

Chapter 10

After the turn of the millennium, tourism boomed throughout the tiny nation of Ricuesta, even into and beyond the little place called C-Town, as well as the Caye. Everything changed. Cruise ships increasingly gave way to plateau cruise crafts as seaweed farming waxed and waned in C-Town and the garbage dumped on Common Senior did likewise. Once tourists started showing up by the shipload looking for something new to see, visits to the Caye increased, and yet the Caye remained something of a secret; indeed, even as it became more visible, there remained a reticence about it. It was a secrecy that emerged as an affective energy throughout the town. This was how I saw it too, from the boat I sat in.

I remember how the fishing boat slid up onto the sandy shore, one of the sole places it could reach land on the Caye. I had been meaning to visit the tourist attraction almost since arriving in C-Town in August, and finally opted for a local tour to bird-watch on and around it. As we approached the island, the tour guide allowed tourists to take control and guide the boat ashore. We could take control because of the new motor that the boat was equipped with and its built-in safety features. I had purposely sought a guide with a Frankl-e motor, called an e-boat, as opposed to some of the others who still used the traditional motors, which plunged into the water and ran on fuel.

The thing about Frankl-e Motors was that they were pretty frank with their customers. Their slogan was, "Fair prices; Fair reliability; Excellent Progress." In other words, they needed patience, support, and ongoing upgrades to continue with their exciting ventures, and residents in C-Town were in full support. Their products weren't entirely reliable, in other words. They were fairly reliable. And also fairly priced. A little slow too. There were competitors, but they weren't readily available or popular. Traditional motors were more reliable, but foul-smelling, toxic, and excessively loud. It was that sound and smell, the dizzying feeling of inhaling exhaust on a humid day, that I'd forgotten about. E-motors were considered to be more environmentally friendly, safe, cheaper in the long run, quiet, and promising, and

C-Town tour guides were on board for these reasons. That, and they appreciated Frankl-e's honesty. I opted for an e-boat not for the calmer experience, but because I wanted to experience the breakdowns—the fair reliability—for which e-motors were well known.

Many farmers, who had been pushed out of seaweed farming due to increased pollution, partially caused by tourism, which wreaked havoc on growth, recommissioned their fishing boats with e-motors and advertised a touring experience that was authentic but modern and relaxing. Other fishers-turned-farmers, who also relied on tourism but could not afford the e-motors and still relied on gas-powered ones, said their boats were truly traditional and touted a nostalgic experience.

It was a difficult decision to make, but I went for the new experience. Nothing was predictable with the new technology, and that was what enticed us. There was a certain secrecy to e-motors, too, then. I'd also heard that, in the event of a breakdown, "breakdown parties" would erupt as a fun way to add to the adventure, and that tour guides even brought rum along for the occasion. They weren't supposed to, but as we glided slowly toward the Caye, it became obvious that they did. Oh, the motor didn't fail. Rather, the tourists' patience surely did, and they started grumbling, so the captain pulled out the bottle and we all had a shot. And, soon after, another. And then another. It helped with the complaining for a little while, but almost led to an accident. Not with the boat, though . . . The problem was, you see, there was no restroom, and one of the tourists was in serious need of one. It forced us to come to a halt.

"Just jump in the water and do it there," his wife suggested.

"But I don't want to get wet!" he cried as he reluctantly peeked at the water to weigh his options. "Ew, fishy."

Shut up and get into the —— water! That was what I was thinking. I didn't say it, but after looking around, it was obvious that everyone else was thinking it too, as the passengers peered through blinding sunlight at the tourist and stared, sighing long deep breaths.

I also wanted to take this trip to walk ashore on the Caye. This was not possible with the old motors because the Wilderness Collective (the official Caye managers) had come up with all sorts of reasons to limit their use around the Caye and forbid their access to it: there was no dock; they were outdated in terms of safety standards; they created pollution; they were noisy and disturbed wildlife. Fishers claimed that the Wilderness Collective was biased, that they were being held back. Whether or not the Wilderness Collective's reasons were sincere, if I wanted to walk ashore on the Caye, I was forced to take an e-boat.

What was more, tourists weren't allowed to "take the wheel" on traditional boats either. It was a far more passive experience, burning fuel—at least, the use of e-motors felt more active, provided that I stayed on course and the autopilot didn't kick on. In any case, there was a certain unreliability and spontaneity to this technology, and that was what tourists banked on when they took the captain's seat.

After all the tourists took turns along the way, and after tipping the tour guide, I was given the handle to get the e-boat onto a small, sandy shore. I was instructed to steer toward an area free of the other boats (only four boats were allowed on the Caye at a time as per regulation mandates) and pushed right onto the shoreline. I looked over my shoulder and watched the water move out of the whirring e-motor, which required very little depth to operate. I then turned back to the shore and let go just as we reached land. The motor immediately quit, lifted above the water's surface, and we slid to a stop on the beach.

The other tourists on the boat, eight in all, hardly waited a second before grabbing their bags and scuffling onto the beach. Most tourists visited the Caye to see the red-footed booby birds—and as I came to realize, the red-footed booby birds made tourists practically wet themselves with excitement. They scrambled onto the shore so quickly that more than one fell into the sand from the boat.

I was there searching for Gregory. I felt a bit unusual about it, given that laughing gulls had become somewhat of a nuisance. I was all too aware of that trouble—that their population was growing exponentially, and that no one knew how to feel about it. They had almost become aligned with lionfish, but it was hardly convincing, as the gulls were an integral part to the environment, even as they destroyed it; and thus, they became a problem for Caye visitors and Caye friends.

Denise told me about the laughing gull colony on Common Caye. No one wanted to do anything about them at first. It was understood that these gulls were finding it more difficult to establish nesting grounds due to increased offshore tourism. The fact that tens of thousands of the birds had made the Caye home, and in a place that no one really wanted to hang out on for extended periods, was just perfect. The only downside to these birds was that they were, as it was put to me, "shitting bastards" that "will shit all over everything you love." This was obviously a reference to a popular meme picturing a bird, usually a pigeon, with the same expression—and it had been around for as long as I could remember. Its silliness helped with its longevity. It did not mean Denise thought it was a laughing matter, though. Not only did these birds pollute the land terribly, but they laughed too much too. They were too loud and too messy. But on the Caye, they could squawk and breed all they wanted and add all the waste they could produce to a big pile of more waste where it really didn't matter, or so it was once thought.

It would soon be discovered that their highly acidic excrement and obnoxious laughter were both killing the Caye and driving out the other birds that tourists wanted to see. It wasn't long before culling entered discussions. After all, after FTCC banded together to address the cruise ship development plans, they later connected with the environmental GPA the Wilderness Collective, and this collective was known for its management and sustainability practices, not to mention its powerful international partners, and the dominance that characterized their movements among its members. It was not long before full oversight of Common Caye was established by the Wilderness Collective. This led to an influx of all sorts of research into the Caye, as well as much talk about bird excrement, and hence, talk about "shitting bastards" that would send you running for cover. It was no wonder that culling entered talk too.

Gregory would have been lumped into this defecating category. And yet, as an aged, categorized, gendered, and anthropomorphized body, Gregory exceeded himself by connecting to others as a potentializing force. For many, it was a memory of the chase down the beach. He consciously, not accidentally, sensed a way to live with others and the drastic changes being experienced in C-Town through the eventfulness of friendship, while living in the precariousness and unpredictability of life in a growing tourist destination.

Memories of Gregory opened into a post-neoliberal "logic" in a place so utterly at pace with constant change and ecotourism development initiatives that he emerged as an emergent relationship within it, working with it as an active force and a lively, active, incommensurable individual in a place where all differences increasingly became individuality with which to move ourselves along and beyond connections made into potential new ones with others. So, there I was on the Caye, looking at birds and looking for them. The smell of sweat and trash, and diapers, and excrement, and salt, and ammonia, and excited anticipation wafted through the air, not to mention the sound of one hundred thousand laughing gulls doing just that. It was a sound that cut right through us on the tour to the Caye. It was a sound no one could ignore, even as we tried to walk along the shore's edge toward a small path that was built for catching sight of various birds. There was no missing the birds like Gregory, or those large, dark fowl flying overhead, often mistaken for drones.

Drones were another matter, but they hadn't been a real issue on the Caye—not yet anyway, not while the gulls dominated. More and more virtual experience was being added to tourism, especially after drones started showing up in droves. Drones were known to swarm around wildlife sanctuaries ever since e-tourism, or virtual tours in the case of the drones, made its appearance in Ricuesta. They took visitors through deep-forest archaeological sites and bird and jaguar habitats on personalized tours that left zero footprint, until they accidentally killed a bird or disturbed them so much they either died or stopped reproducing. The drones proliferated to such a degree that avid bird-watchers began including them on their species lists and attempting to name the make, such as Dawson's Destination Drones and E-Flight, designed and trademarked by Sea Green Vacations. There was even talk about undersea drone tours geared to the reefs, as well as diving the sunken cruise ship in the works. Although in limited numbers, drones were coming in and were seen hovering over the Caye too. I could hear their buzzing and see them hovering about when I arrived there. But what I saw that day on the Caye wasn't a drone. It was a magnificent frigatebird.

It was common for tourists to mistake magnificent frigatebirds for drones, but I could make out the neck against the clear, blue sky. It was shocking that other tourists couldn't do the same. That didn't mean I recognized the species before an avid birder told me, however. They had just gotten onto the beach when I pointed up at one as it quietly flew overhead. "What is that?" I asked.

It was my first visit to the Caye, and I was taken away by how calm yet simultaneously busy the place felt. I felt so alone, yet so taken away. My memory of the magnificent frigatebird is the first thing that comes to mind when I think of that place now. I'll never forget the look of direction embodied in the contours of its midflight neck, its reach into the blue. It was leaving. *Where is it going?* I wondered.

There was no forgetting the scattered display of cement, broken tree branches, and metal along the shorelines of the Caye either. I hopped across them to snap pictures, and nearly twisted an ankle. They were the remnants of old discarded foundations informing haphazard stepping stones across the shoreline—informing its history.[1] They informed, in other words, the ongoing transduction of one milieu into another in time. It was felt as the experience of bodies and trash in passing, its source of energy, its reach into the past, its self-relation and "self-complication," a place where time was spatialized as a dimension, and an experience rather than an emplacement.[2] It left me reeling and tripping, searching. The place itself passed through itself and was experienced as twisting, open-ended connections reaching across endless bodies and things folding in and through multiple dimensions of space and time, and it took me into it and away to the sound of laughter.[3]

While on the Caye, everyone spoke in whispers. That was insanely weird given the gulls. It was as if the tourists didn't want to draw attention to themselves. But it could have been the immense summer heat and the smell of the Caye that kept the visitors from deeper inhalation. After I mustered up the courage, I drew a deep breath and called for Gregory.

"Gregory!" I announced at the top of my lungs. The air warmed my chest as I pulled it in and nearly coughed it back out.

"Greeeegoryyyyy!"

"Gregory?"

My fellow tourists were staring. The confusion was palpable. Unlike the others, though, my tour guide knew what I was up to. As I paused to take another breath, he relayed information and, with my lips still parted, I froze immediately. He said that the bird was dead. The word struck through me. *How can he know that for certain?* I thought. I still didn't call Gregory's name for the rest of the trip.

Notes

1. Interestingly, across the planet some of the basic materials for urban development, including metal and cement—albeit specially formulated variants—are being used as support structures on which to regrow corals lost to climate change and other destructive forces.
2. See Massumi (2002: 13–14) for his discussion on sensation as "immediate self-complication."
3. "The notion of the milieu is not unitary: not only does the living thing continually pass from one milieu to another, but the milieus pass into one another; they are essentially communicating…" (Deleuze and Guattari 1987: 313).

Figure 11.1. "Ravenous Appetites." © Ines Taccone

Chapter 11

Both FTCC and FLBCC banded together on the cusp of transition between traditional cruise ship tourism and plateau cruising. It happened when cruise tourism reached new heights of aggravation for locals dealing with its intensification. It was as if matters couldn't get any worse, but they were far from over. It was no surprise, then, that a development company showed up on the shores of Common Caye and Common Senior. And those who loved that land-and-sea assemblage knew it was yet another beginning of a whole lot more mess.

Given that I arrived in C-Town years after the Caye became a contentious wilderness, I had to do some digging.

"We are relatively well-established, being about a decade old," a representative—and one of the original members—of FTCC told me in an interview. I'd interviewed FTCC friends as part of the report I had to write for Rita and CORALREEF. Rita thought that it was a good idea because FTCC members worked with CORALREEF on several projects. That also gave me the chance to ask about their work with the Caye.

"That explains your name," I said.

I had been wondering about their name since hearing it, because a four-letter acronym for a GPA was unheard of unless they were established before sixth-wave NGO-ization—at which time most NGOs became GPAs and their numbers rose exponentially. Even individuals GPA-ized themselves. With the web-based integration of all GPAs worldwide, nothing shorter than thirteen-letter acronyms were left for the taking.

"Well, no. Given that we were not initially a GPA, and certainly not official, we did not have to abide by those rules," the representative responded. "Now, we piggyback off of another GPA; otherwise, we toyed with the name POTSATEOTS_123 until we discovered our relationship with the Wilderness Collective."

FTCC did not initially intend to be formally registered as a GPA, which was required if they were going to partner with other associations and create formal agreements. By registering, they agreed to oversight by yet another association, which was established to coordinate enviro-GPA activities. This was especially important with respect to environmental matters as associations normally partnered to help manage environmental sites. FTCC wanted no part in managing the Caye, however, even as this became an expectation. They formed the group and referred to themselves as "friends" with the intention to complicate the redevelopment of their beloved caye. They understood that development and change were integral to its existence and, through the unofficial collection of members, made such a process unavoidable—that the Caye's emergence was to continue with the emergence of relations, not despite them. If FTCC members were to have no part in management, they also felt no need to formally establish themselves as a GPA.

It was local fishers, many of whom turned to tour-guiding, who were the first to start up FLBCC, which was initially an informal collective. They did not loathe the tourism industry, but its physical immensity. Ever since Edison Energy emerged as the major competitor against Frankl-e Motors, followed by a number of smaller corporations in its wake, including Diesim and General Electric Motors, "sustainable" transportation expanded to every industry and to every corner of the earth. That was when a new form of tourism came to Ricuesta. It made travel cheaper than ever and supposedly eco-friendly too. This expanded the extent and intensity of the tourism industry, especially in partnership with the environmental sector. It was all very promising and immense.

That was until the batteries started to be lost at sea every now and again. E-boats were very quiet, and coupled with their built-in, fairly reliable navigation systems, collisions became a habitual occurrence. And when they went down, the batteries went with them, and that was where they stayed. GPAs, such as CORALREEF and CAPTLUPSITALMITY&CODS, popped up to collect the wreckage, although the damage was often already done. Luckily for tourists, the e-motors' autopilot systems made them appear "safer" to handle, and this was what allowed tourists to take the reins and navigate the boats when visiting Common Caye.

People and animals had grown fond of these e-motors because of their quiet humming, which bothered no person or fish. Even cruise ships partially switched to these electric motors. Before their popularity, the shoreline was a bustling zoo of rumbling boats, growing louder and louder each season. With e-motors, the beach was again a more peaceful shoreline, that is, until ships came in and dumped their passengers on the shores. The motors, although initially more expensive, were cheap to run, at least for the cruise ships, which, upon switching to e-motors, "gassed up" in North City—Ricuesta's largest coastal city. The term "gassed up" was used instead of "recharged" because, although efforts were made, Ricuesta did not have the means to invest heavily in sustainable energy production. Electricity was generated from burning fuel, in other words. And one source was biofuel.

Support for biofuels was made into a growing popular vision for tourism-oriented coastal nations due to spreading success stories around the globe. Soon, North City was made into a major hub for cruise ships to "refuel" for cheap. First, sustainable sources of energy were used, and when the demands quickly exceeded the availability, they turned to importing, as well as a reliance on fossil fuels. There was ongoing wind turbine and dam project talk while I was in C-Town, but alas, it was a very expensive endeavor and was historically proven to be not only unreliable but also highly disruptive. This is not to mention that Ricuesta was in line to receive help from companies and nations intent on using cleaner methods of transportation, but that help was much too slow. In the meantime, they recharged their "sustainable" transportation systems at the city's edge.

And this was the beginning of the end for cruise ship tourism. With the advent of cheaper than ever travel, and given what these motors were capable of moving, the new plateau cruising industry appeared practically out of nowhere with its huge tourist resorts atop equally huge, flat, glass crafts—resort nations, some say—that floated on top of the water with their many e-motors located under the craft. There, the motors dispersed water, handled waves by continually cutting and supplying power to different motors, and allowed these huge floating cities to glide along the shore's edge, silently and quickly, appearing almost like icebergs in the distance, while promoting eco-friendly travel.

Their immensity could not be missed. To conserve energy, the floating resorts were built to link together, which allowed them to cut multiple motors. Once connected, they extended out over the water, bubbling along, so that something that looked and was the size of a small nation literally showed up on seaside communities' doorsteps and suddenly made the seaside inland. Part of the appeal was that plateau cruise crafts did not need deep water, not even much more than half a foot. Then, as if this was not enough, they could get up on the shore with the use of their massive, folding glass platforms, which ran the length of the craft and could extend beyond it to rest on land. When these crafts approached the shore to either rest or recharge, the waterfront was totally gone, the seascape invisible, the water turned to glass.

Although these crafts were—relatively speaking—efficient, and although they were equipped with solar panels to recharge, none of it was enough to keep up with their energy needs. As such, they moved along the shoreline to recharge as needed. Recharging was in serious demand when these cruises traveled down the coast. The floating resorts, thus, stopped at major hubs before moving out to remote places to rest and recharge for days or weeks with their solar panels. If no further stations existed close enough to continue their journeys without a need for extended stays to recharge by way of the sun, they headed back north. The industry goal was to move out further and further. In Ricuesta, they got as far as C-Town.

To make matters more complicated, given that plateau cruises were equipped with glass floors, they were hugely popular wherever there were good, easily accessible, nearshore undersea views, especially near coral reefs. As a result, the ships did not consistently beach themselves in C-Town. C-Town had no recharge station. Instead, they remained underway, basking in the sun, and completely covering over major fishing spots. All that residents—and whatever was left of the land-loving tourists—saw at that point was a wall of plateau cruises out on the water at any given time, with none of the consumerism that tourism was supposed to bring. To deal with this matter, plateau cruises were mandated to take turns coming to shore and allowing tourists off for several hours. In this way, crafts were individually distributed down the shoreline while others waited their turn. Then, with no warning, they quietly drifted away.

With the rise of e-motors came a need for more and more energy, as well as better storage systems. Ricuesta could hardly keep up. They had no choice but to import much of their energy. To add to this, tourism entailed more garbage in the waters, and the habit had a nasty way of showing up on shorelines almost anywhere. As on-land tourism waned in C-Town, leaving huge cement resort buildings across the waterfront underutilized, and the cruising industry grew, more garbage and even more pollution than ever entered the waters. It had tremendously negative effects on seaweed farming in C-Town as well. Many a farmer was pushed into agriculture in support of e-tourism, or more into tour-guiding, there with their e-boats. So, when a development company proposed to redevelop Common Senior, everyone knew it was not because the cruise industry required a place for their vessels to come to shore. They knew it was only a matter of time before the site became home to a recharging station, bringing with it the intensification of travel and energy production in Ricuesta.

C-Town was caught in the middle of it, literally and figuratively. It must have been hard to oppose a development scheme that would bring new employment opportunities, but there was something about that Caye, maybe even something about that bird that we used to know.

FTCC members just wanted to be let be, and by that, they wanted the same for the Caye. That was why they banded together. Some even saw the FLBCC and its alliances as a resurgence of the former fishing cooperative that sunk once mass tourism showed up in C-Town. But this assemblage of friends was not a replication of any group. They say that those who fell for the Caye were drawn to it repeatedly, and they explicitly called it "love." In a newsletter by FTCC entitled "Falling for Common Caye," a friend—and representative of the Wilderness Collective—wrote,

> People find the place and fall in love with it. It makes us greater than ourselves. It's magical. The place has a magic to it . . . We have had such a supportive membership despite the ups and downs . . . But I think the takeaway point is the feel of it. It's the only place like this in the entire world . . . Members can't describe it; it's a feeling. They just can't give up.

The emotion couldn't be explained even as it returned over and over. It could have been considered a practice of what Donna Haraway called "becoming with," but here, the magic wasn't in the fact that lovers had come together. The magic was that despite such a desire to crawl into their beloved and remain in the warmth and unity of love, these friends wanted no part in management, and thus, had to be open to letting it go.

In other words, there was more than an intention to preserve the Caye. FTCC members opposed the imposition of a new direction in their relationships with the Caye because impositions would render the various relationships with Common Caye moot, and outside its future. The issue wasn't that a new direction was imposed on the Caye. It wasn't in whether the Caye had a choice in the matter either. FTCC members did not argue that their love for the relatively new, natural-synthetic landscape earned them a right to speak for it. Instead, FTCC members faced a conundrum. Things were taking a turn again. That's the thing about any relationship; it changes, as does the feeling. The very fact that anyone fell in love with the Caye in the first place attested to the fact that to fall for it was to change how they felt about a mountain of trash. The new relationship entailed movement away from a previous one, and that was what it was to love the Caye, to let it go, to fall for it. The proposal for a cruise ship enterprise to attach to the Caye did not threaten FTCC; it didn't surprise anyone; but it did change how they felt the moment the development company showed up. That was the tragedy. They had to start over, with all the mess of letting go, and of working out something else. If Caye-lovers simply shrugged their shoulders at the proposal, their relationship would not continue. In fact, had they shrugged their shoulders in apathy, the love would have never existed. But it did exist, and with it so did the problem of letting go. And that was what they struggled with as they faced the new proposed development of their wildly synthetic, emergent landscape. And hence, "shit hit the fan."

Falling for the Caye was an ineluctable caprice, a touch that created uncertainty, and was therefore a desire and a regard across difference, without the ability to expend that difference; and consequently, it also magically produced differences in desires, memories, and more. It was an uncertain desire, to fall for the Caye. It might be that this is what happened when Gregory ran at his visitors with everything that made him excited to do so, and they ran back, often with camera in hand.

Perhaps it's no coincidence that friends tend to mirror each other in conversation.

Chapter 12

When I remember Gregory running at those passing by, and them running back, it comes as a moving image existing in a single moment. Nothing is connected before or after this memory. It appears whole. And yet, as that experience is repeated, it seeps into stories. Through each moment of remembering, our connections, pasts, and relationships emerge and diverge, again and again. It is another way we travel.

FTCC emerged with an image of opposition by which they would be remembered. With this image, they cast out to the world a nostalgic call to the future. At the same time, it allowed them to cast out lifelines. They set up a website, prepared newsletters, held meetings, spread their "word" through tourists, connected with environmental workers at home and abroad, and reached out into the world through social media. They emerged as an enticing site of potential, as one member wrote in a newsletter:

"Here are coral nurseries. This is Nature's Last Stand. Here is the natural beauty of perseverance but treated like it doesn't matter!"

With this image, nonprofits from other parts of the world contacted FTCC. One happened to be a local advocacy group located in Toronto, Ontario. As luck would have it, one of the members was an avid sailor who discovered Common Caye and its predicament while on a trip to C-Town. Upon hearing the news, this visiting sailor put the newly formed advocacy group in touch with her connections, and it was then that FTCC came up with their slogan, "We are the Caye," and received help in producing newsletters alongside awareness-raising strategies.

It was, in part, due to their partnership with the Toronto-based advocacy group that FTCC developed their approach to blocking the redevelopment of Common Caye. With connections to a growing membership base, as well as major enviro-GPAs, including Audubon Societies from different parts of the globe, and later, the Wilderness Collective, FTCC's membership, ties, and appearance grew immensely. In fact, as one representative said, just months after FTCC began raising awareness, their online presence became endless. With this, FTCC was able to gather enough clout and support to appear as a site of required connection for all things Caye-related. So serious were their demands that threats to tear the whole Caye up if these demands were not taken seriously, as horrific and contradictory as that sounded coming from friends, were made. As mentioned already, their slogan was, "We are the Caye," and it was promoted as such for a reason.

Given FTCC's refusal to register as an official GPA and manage the Caye, some groups, especially the cruise company's officials, argued that FTCC's concerns were illegitimate and purposeless. Contention within FTCC ensued, with some arguing in favor of ignoring the demand to register—claiming that it was unnecessary to be officially recognized as such an association—while others argued for registering and moving on. The ongoing back-and-forth debate was, some feared, taking too much time at a critical moment. Then, unbeknownst to FTCC members, the legitimately established GPA, "The Wilderness Collective of Undifferentiated and Indifferent Animal Lovers of Ricuesta and Beyond Who are Care Professionals for All Animals and Some Other Species on an Ad Hoc Basis," also known as WCUIALRBWCPAASOSAHB, or the "Wilderness Collective" for short, registered FTCC as a chapter under itself, which automatically registered the rogue organization by default. Wilderness Collective members were already known for their domination over environmental matters and partnerships, but this was a new high for them. FTCC were simply volunteered into a knotted relationship with the GPA.[1]

As a result, two groups under the same name emerged. FTCC became a formal, organized representation of the Caye, as well as an informal friend, there to complicate matters. Their name appeared under two websites with separate memberships for each. The membership dispersed over the two identical names, and this led to much confusion. I can only imagine the conversations:

Wilderness Collective member: I'm heading out to the FTCC meeting tonight at town hall. Are you going?
Informal member: Isn't it an outdoor meeting, by the old pier?
Wilderness Collective member: I thought it said town hall.
Informal member: Oh, I guess that's where then. See you there?
. . . Later . . .
Informal member: You tricked me into coming here to up your formal membership numbers!
Wilderness Collective member: I didn't even know there was another meeting!

I even discovered that, in some instances, members of both FTCC groups intentionally attended each other's meetings. The thought of two rival groups with the same name meeting in the same space blew me away. I imagine the conversations occurring along these lines:

Wilderness Collective member: So, we are in agreement with the Wilderness Collective's initiative to construct a path on the Caye for tourist use, yes?
Informal member: If FTCC is okay with it, so am I.
Wilderness Collective member: Yes, FTCC is in agreement.
Informal member: Okay, then I agree.
Wilderness Collective member: Wait, which FTCC are you a member of?
Informal member: The proper one.
Wilderness Collective member: The proper one? Sorry, I'm a new member, and I'm still confused by the divide.
Informal member: I'll let you in on how it's done . . . Anything the Wilderness Collective's FTCC-er branch promotes, we push back on because they hijacked our name and group. It's why we dropped the T from their name.
Wilderness Collective member: Why would you do that?
Informal member: Why do you think?
Wilderness Collective member: Well, as a matter of fact, I am a Wilderness Collective member, and in our defense, if you'd just partner already, we could move past this childishness and make some real progress.
Informal member: This is why we dropped the T.
Wilderness Collective member: You are clearly unhappy with the management strategy, so I'll take it you aren't for the path.
Informal member: But I like the path idea.
Wilderness Collective member: So, you are in agreement with FTCC?
Informal member: Yes, I am a member of FTCC.
Wilderness Collective member: Which one?
Informal member: The one that agrees?
Wilderness Collective member: That's the one under the Wilderness Collective.
Informal member: Oh, no no no. I was just talking to another FTCC member, and they verified that FTCC does not agree with FTCC.
Wilderness Collective member: So, no to the path?
Informal member: No.
Wilderness Collective member: So, you agree with us then?
Informal member: No.

Eventually, FTCC was forced to communicate and link their websites. But as the townspeople and FTCC's many members realized time and again, the confusion had only just begun. And as another anthropologist, Anna Tsing, has argued, confusion can be more productive than clarity. This is partly because confusion is, in some ways, desirable. After all, confusion offered a sense of consistency in freneticism that did not necessarily feel frenetic, but promising. As for FTCC, when its members consolidated and formalized their membership, it did not mean that their friendship was made easy or less confusing. Rather, it was funneled into a new, surprise partnership with the Wilderness Collective, which provided the validation they needed to be taken seriously by the cruise company. Yet, more than confusion, something else emerged from the fact that with partnership, friendships were bound to happen, and friendships were complicated, fickle, and somewhat unscripted relationships. Friends really got on one's nerves.[2]

When FTCC discovered that their name and image had been taken over by the Wilderness Collective, FTCC representatives were upset that decisions had been made on their behalf. But the group itself did not function on agreement. Friends of the informal FTCC itself were divided over the Wilderness Collective's strategy. Some felt like they could work with the collective even if they didn't like it. Others didn't. Some were indifferent. I wasn't there at the time that this happened, and upon discovering it, I thought it was a very clever move. But then again, the Caye and its friends were not friends I chose to remember. Trash lived vicariously through mine and Gregory's relationship. Our relationship, in its dispersal, opened the possibility for memory where it otherwise would never have emerged.

It is all very confusing, friendship.

Then, as if these rivaling friendships weren't enough, FTCC was made up of residents, environmentalists, volunteers, and advocates from home and abroad. Of these participants, and making up their own group within it, were fisher families who allied with FLBCC, which included families who'd relatively recently moved to C-Town. The number of fishers involved was part of the reason that C-Towners saw the movement as a sign of reemergence of the former fishers' cooperative—one that had historically managed conflicts between fishing families, albeit not always successfully—yet it was also entirely new. After all, the Caye drew enduring returns by way of a touch and a pull in ways never felt before.

With this newness, strange friendships emerged. Despite family loyalties, there were no stark lines where certain fisher families fell on one side of the divide or the other. Not only was this shocking for those who suddenly found themselves potentially opposed to members of their own families, while allied with those with whom they once had tense relations, they were also partnered with all families and more under the Wilderness Collective, leading many to accuse the GPA of showing up, causing turmoil for families, and dominating the Caye.

Then, as other town members, environmentalists, and advocates broke off onto either side of the divide or both, they found themselves caught up in fisher conflicts as well. Even non-fisher members began accusing the Wilderness Collective of conflict. Those who had agreed to formalize FTCC, on the other hand, argued that the Caye was their main concern—to which those who preferred informality responded in a nutshell with, "Yes, exactly." Thus, confusion proliferated.

Then rumors spread too. Rumors about ties to e-tourism companies and the cruise industry popped up after the Wilderness Collective poached FTCC. That just added to the uncertainty. Before consolidation, no one knew what to believe, or which meetings to attend. After consolidation, no one knew who to believe or which meetings to attend either. Yet it was due to the confusion that FTCC eventually agreed to the partnership. But it didn't end confusion. It expended it into further, productive, albeit complicated, friendships, to the point that people were completely lost, not knowing what to think, which way to turn, or what could possibly happen next.

Notes

1. For more on knots, and "that which is hiding in relations," see Yusoff (2013: 218).
2. Writing on productive confusion, using an example through which a forest was saved, Anna Tsing writes: "Collaboration was not consensus making but rather an opening for productive confusion" (2005: 247).

Chapter 13

FTCC was made a partner and then the partnerships reached back into the Caye where the Wilderness Collective also began intense scientific research. That was when excrement became a big problem. The issue with the gulls was: instead of enjoying the already established Caye environment, they were ruining it with their presence. And just as the gulls were ruining what had emerged as a miraculous network of trash, trees, and wildlife with their presence, the same could be said about the infamous lionfish. These fish, known for their ravenous appetites, had invaded the Caribbean Sea decades earlier—and sort of like the gulls, they were also aligned with ruining the sea with their presence. This was because lionfish did not promote connections; they ate them. And what they were eating were vulnerable sea life-forms that fishers, other environmental workers, and even the vulnerable life-forms themselves were trying to promote; yet the lionfish continued eating without finding a way to reconnect, that is, before fishers and GPAs acknowledged the problem and reached out. They reconnected, not surprisingly, by having their members—particularly tourists—eat lionfish.

It was excessive. And it became a way to love and hate lionfish at the same time. "I don't hate lionfish," one volunteer said over coffee in an interview. "There are broader things to consider." The word "things" was often used by young volunteer workers, and no one expected them to find anything more specific to say. How does one put a word to a feeling?

As for the gulls, excess was the issue. They did not maintain connections either. They got ahead of them as they buried the Caye as an object and foundation for their own reproduction; and the Caye literally lapped it up. The Wilderness Collective would not fall behind.[1]

Because of this fiery struggle and because of FTCC's partnerships, which brought about their presence on the world GPA stage, the Ricuestian government made the Caye into an "artificial natural habitat" and a National Park under the management of the Fisheries Department, in partnership with the Wilderness Collective and comanaged with FTCC. The image of the Caye as a self-renewing, secret wilderness then shifted to a new image of "Nature's Last Stand," which could then be preserved as a park. As a result, the development company eyeing the Caye still received permission to develop their plan, but they would have to go around Common Caye to Common Senior. What was more, they would have to avoid the ridgelike structure extending under the water. The Caye, then, avoided a near-total catastrophe, or so it was imagined by friends. At the time of my visit, development hadn't yet commenced on Common Senior; and no one knew if it would, given the mandated detour.

FTCC, following the lead of their connections, would not represent themselves as managers of the Caye, even as they agreed to comanagement strategies in partnership with the Wilderness Collective. The park was so obviously unnatural and yet totally wild, and everyone knew that Common Caye had received its name in a touch that drew enduring returns. FTCC, the trash heap, the gulls, and Gregory, among others, would have to contend with the wilder and wilder wildernesses and contradictions that that would entail. Thus, excess and reconnection took center stage and became a site of concern and opportunity.

At the same time, there was something else to that caye, and its friends drew attention to it: that in a space where connections amplified, expanded, and proliferated, bonds were stretched thin.

Note

1. "The ability to mime, and mime well, in other words, is the capacity to Other" (Taussig 1993: 19).

Figure 14.1. Cormorants nesting among skeletal trees, Toronto. © Ines Taccone

Chapter 14

That day I visited the Caye with a tour group, my e-boat was one of five on the Caye's shore. In addition, I had to wait for a specific time before my tour guide departed. That schedule of limitations called attention to how struggles surrounding the Caye affected fishers' and tour guides' use of it and its waters. They, too, were limited. Meanwhile, the Wilderness Collective received funding for a project to transform the Caye into a public attraction, not only for bird lovers but for snorkelers and divers.

It was already known at that point that the new, somewhat hidden environment beneath the water's surface around the Caye was acting as a source for coral regrowth. In partnership with CORALREEF, the Wilderness Collective embarked on a project to increase the extent of coral thriving on and across scattered debris around the Caye. Together, they would promote the regrowth around the Caye, as well as around the sunken cruise ship site(s), which would be located close enough to include both the Caye and the sunken ship on the same tour. Their intention was to partner with other environmental associations involved in reef revitalization as well and use cement pretzels—the latest development in reef revitalization—to construct coral gardens. It turned out that the twisted platform worked better. In this way, the intention was to add to the revitalization already taking place, as well as to set up viable, permanent tourist attractions on and around Common Caye itself. Research, thus, went deeper into the Caye, and into the surrounding waters.

Denise was involved too with her "one more" volunteer ecotourists. I'd asked her if she could inquire with as many tourists as possible about Gregory's whereabouts. Maybe, just maybe, someone had run into him. I never heard back. Denise was so busy at work that she probably didn't have time to inquire about a bird. The Caye emerged out of self-revitalization—so it was imagined—and it was imperative for all its friends to keep up with, even get ahead of, its own redoing.[1]

It had been discovered, through research with volunteer ecotourists and various groups and researchers, that the Caye was very good at holding itself together. Toxicity levels, though, they found, did have a negative effect on fish and mangroves in some areas. Although temperatures were down in this microclimate, toxins leached out, and seeped into the surrounding environment. There was a warning, at one point, that certain areas, which were called "burn zones," could poison the unsuspecting swimmer. It was believed that toxicity was released as a result of storms. No one wanted to paint Common Caye itself as toxic. Instead, it was damaged, and these damaged areas were individual sites in need of further revitalization. More mangroves would have to be planted where they thinned or died. Burn zones would need to be traced back to sites of leakage and cleaned up. With regard to the debris that was scattered across the seafloor after storms ripped parts of the Caye apart, some of it would need to be moved, but also, some of it was already serving as areas to aid in coral revitalization. The plan was to expand those sites. And then on the Caye itself, there was the problem of acidic bird waste and overpopulation pushing out other forms of wildlife. As a result of all these new sites of revitalization reaching for more, volunteers came and went, researchers and grad students showed up in swarms, and various GPAs offered support in relation to particular projects, areas, objects, species, or groups.

Denise was busy with organizing tourist outings, collecting species lists, coordinating meetings, and volunteering her time to help with the annual Key to the Caye Festival, which became a branch of the annual Lionfish Festival in C-Town. She even had to deal with the new crisis of ecotourist luggage. Smart luggage—with its ability to follow its owner anywhere with its small but durable tires and software programmed to track its owner's cell—had the unfortunate habit of escaping from cabanas or even off plateau cruises and plowing through protected forest areas or ending up in the water in its efforts to reunite with its absent owner. So determined were these bugged-out suitcases that it was common to hear them banging against the walls of cabanas and hotel rooms, and then making an escape when housekeeping opened the door. Stories about runaway luggage filled with drugs rolling through the jungle or following unsuspecting tourists through the airport ran rampant. On two separate occasions, once while visiting a ruin and another time on the beach, I ran into tourists, who, after feeling a familiar sensation of something behind them, turned around to see their suitcase sitting there, seemingly content, like a pet. It somehow became Denise's job to chase these escaped suitcases down, while covering all her other responsibilities.

Denise had been trained on how to override the program and redirect the luggage. She didn't like it. "With this back?" she would say, even though she didn't have to actually carry the luggage, just redirect it. The time and energy required were not worth the trouble, and so with her newly pocketed resources to override the program, she hired someone else to chase the suitcases for her. Amid the confusions, movements, and turnovers with respect to volunteers and partnerships, and agreements and disagreements, which Denise had to deal with constantly, the thought of having to deal with more confusing matters, like Gregory—especially after she'd just offloaded one concern—was likely too much. And that was exactly what the laughing gulls did. They did too much.

I didn't doubt for a second that Gregory added his fair share of mess to the mix. After all, one fond memory that locals shared was how tourists, after laughing with him, feeding him, and letting him walk on their laps, got crapped on—sometimes in their hair too. It must have been a sight to see.

I could imagine Gregory returning to his nest daily during the time that the new FTCC began their attempt at managing the gulls' nesting habits, by disrupting their grounds and later trying to lure the gull colony over to another island several miles away. But the birds returned. I imagined Gregory never left. That he had too many memories on the Caye to leave, and too much tourist pomposity to take down a notch. Or that the Caye, in its own way, asked him to stay. That he enjoyed the sunsets, the proximity to the beach where he ate from the hands of passersby, where he may have laid eggs—yes, possibly—where he may have hunted, watched trees grow, picnicked, laughed, and fished with locals or tourists on their e-boats. But all that changed for him as the increasingly wilder-nesses of C-Town continued to unfold.

The Bird–Caye–Advocates friendships began with life-forms wandering into urban sprawl's excesses. Out of this wandering emerged new identities, alliances, proposals, legislations, associations, dreams, and stories—all of which were beginnings that began with a feeling and a fall.

The Caye remained productive of secrecy through all this. It was full of surprises. Magical in a sense. Falling for it generated a kind of magic. C-Town was full of these stories. Stories went around about people who were known to run into money just when they needed it most, just when they fell ill, or when a family member's e-boat broke down, as if someone was watching over them. There were stories of big breaks but also loss, a whole lot of aggravation and a whole lot of pain caused by those loved most. No one knew where alliances truly lay, not when they were always changing. And they had to keep shifting because no one knew where they were going. So, secrecy abounded, as did whispers and speculations.

Gregory had secrets too. He wouldn't tell me where he'd gone, if he'd gone.

Denise said that her dealings within her GPA network were also somewhat discreet. They needed to be, as she needed to renegotiate her contract every few months. Between renewals, Denise learned to pocket potential sites of reconnection to whip out when needed. When it was discovered that lionfish were breeding up a storm around the Caye, no problem; Denise had anticipated something along those lines. She hadn't told the others what she was doing, what she was anticipating. She didn't even really know what it would be. She worked. And she waited. Then, when it was time, she already had viable connections and friendships in hand. She had learned about lionfish, and corals, and became involved with members associated with the C-Town Lionfish festivals and Lobster festivals.

Denise liked coral; she liked the Caye too. But she had mouths to feed, debts to pay off, family members to help. What she really needed was money, and if she had to say that she totally cared for the Caye and for corals, then she'd say she was hopelessly and madly in love with them, head over heels if need be. She didn't tell anyone this. She especially didn't say that she was helping to provide one of her kids with the funds he needed to get involved with seaweed farming, a still viable practice in C-Town, even if rocky at times. No one needed to know about that. All they needed to know was that she could do a good job. And she could. Her ties to the community were also gold, especially since future work was geared toward further engagement with the local community.

Denise said that the unpredictability of the contracts was worth it. She went with it. Fell into it. She wanted a more secure job, but she had to contend with other secret and not-so-secret relationships as well. Denise's colleague, Steve, had a position in the Wilderness Collective that was much higher paying than hers. He was well known, well connected, well educated. There was more to it too. He had a connection to another very large international wildlife advocacy association from which support came pouring in. Everyone knew that his "connection" to that association was more than professional, but no one said anything. If two animal lovers wanted to work together and it worked for them and the animals they loved, then who was anyone else to say anything?

That is, until the relationship fell apart. The two were known to disagree hotly over their management initiatives. One argued for complete relocation of the gulls, the other against it, for instance. When the relationship disintegrated, the funds went with it. Rumor had it that it was because their relationship ended that the disagreements intensified, but on the other hand, it was possible that their disagreements over the birds they loved created animosity in their more-than-professional relationship. In any case, the funds dried up and Denise almost lost her job. But luck was on her side.

Denise, after all, had her secrets too. When fishers started noting a decline in fish around the Caye, her colleagues didn't listen to them. Denise did. She participated in every activity and with every group that dealt with waning fish stocks. She helped with organizing C-Town's annual Lionfish Festival, for instance. She did the same with fishing tournaments, and other festivals.

At one festival, she connected with various environmental workers and even attended a workshop on how to fillet a lionfish and discard or destroy the venomous parts to render the fish edible. The fish became a world-renowned delicacy when tourists began traveling to different parts of the Americas on environmentally conscious vacations. Lionfish fishing had become so lucrative that some said that was how the lionfish had actually been wiped out around Florida and were later "replanted."

The tourists eating the lionfish loved the risk. When asked why they chose lionfish they said they "risk it for nature." But it didn't take much to figure out that they did it for the feel of it. However unlikely it was to accidentally consume the wrong or undercooked part of the fish, it was still to flirt with death and enjoy it.

What was more, many of these lionfish seekers came straight off those plateau cruise crafts. I, for one, loved the glass floors on plateau cruise crafts because the entire floorspace was transparent both inside and out, except where the motors were located. Those floors gave the impression of living within the edges of a watery world, like sitting on the edge of the pool, feet dipped in. Tourists felt that movement into a different world on those crafts. "You're in it. You are there, even if you're above it" was commonly heard from tourists. The in-between feeling mimicked the feeling of slowing down and gathering intensity at the same time. Cruisers felt the pull of the waves when standing on the floors. They could see into the water when the floors lit up, and into the darkness when the lights were off. The floor was cool to the touch. Less than a desire for what they didn't have, then, there emerged a desire for more of what they did have. By the time the ships stopped cruising, passengers were just itching to dive deeper into the water. Most enjoyed swimming and diving in the sea once the crafts stopped to recharge. They entered new worlds through a slowed immersion. They entered a feel for the movement over the course of vacationing, dispersing it over time and space, getting a feel for and getting over the overstimulation of uncertainty.

"How was your dive?" I asked a diver one afternoon, after she surfaced from her first ever dive.

"A really very chill experience!" she replied.

"Blow your mind, never experienced anything like it in your life, amazing?" I followed up, a little too facetiously.

I asked this question because I knew the dive wasn't exactly new. There was a week's worth of glass floor experiences, not to mention practice in a pool, then in shallow water. It was all still "very chill" though. It was the production of composure in the midst of chaos, the making of the chill, calm traveler—a transferable skill.

Tourists were taken to the edge of experience and gradually into a new one—that is, except where cell phone connection faltered. Then, there was no transition, just utter disconnection. Periods of outage. Those were common earlier in the century. And it drove tourists to distraction. Before C-Town constructed the new transmission towers, as soon as ships arrived, frantic tourists poured off into hubs on land to connect to Wi-Fi. They frantically searched for power and connection. They ran into restaurants, arms flailing, flagging down their travel companions who'd fallen behind: "This place got Wi-Fi!! Come! Quick!" Complete disconnection led to a near nervous breakdown, with people guzzling alcohol in attempts to pass the time, or even opting to pay exorbitant fees for limited, slow connection while at sea.

This was not to say there wasn't a feeling of serenity in disconnection. It led the bored tourist to talk to complete strangers. But there was only so much of that they could take, hence the rush to shore, hence their panic being too great for them to notice the laughter and tears from those working on the ship when the connection unexpectedly cut out. Tourists were likely still too intoxicated to feel the sadness of separation, and alcohol especially, in addition to caffeine, helped them push their bodies further. It helped them to push their limits, to push past the feeling, stay up later, talk longer, eat more or not at all, get higher, stay awake longer, push their burned and bitten bodies further than they otherwise would or could. The cruise itself didn't do this for tourists. It took them to the edge and prepared them for the dive. With that being the case, tourists could add excitement at a rate of their choosing, with alcohol, lovemaking, the frenzied consumption of tour after tour, riskier, deeper adventures in pursuit of the feeling, diving past the limit, diving with lionfish and eating them.

Diving also taught tourists how to attend to the task. It taught a method of comportment through the shock of difference: how to breathe, how deep, when, how to gauge the surroundings or any equipment they may or may not have, how to move, when, where, how quickly, how to listen to their own heartbeat, to their own breath, to sounds unrecognized, how to handle mounting pressure, how to prevent their eyeballs from exploding. How to keep moving. Under the water their sense of self came undone and new relations with radically other life under the sea emerged. Divers even found a way to relate to shades of light, and all this returned to the feeling of feeling differently. They bought into this. They invested into it. They bought a diving board from which to throw themselves into the blue.

Cruising didn't push tourist bodies over the edge of the familiar. It took them to the edge and dipped their toes into more. At one time, tourists had to deal with periods of disconnection from the online world, but with new cell transmissions, we didn't even need to deal with that anymore if we didn't want to. After that, the tourist stride took on a different quality as their adventures stretched out over space and time. Cruises took time, which maintained a sense of linear progression and left enough time to adapt, breathe—like working on a base tan before embarking on a tropical vacation. It was, then, up to tourists to push their limits, if and when they chose, if and only when they were ready to venture out on their own. They made that choice. They invested into it or pretended to be surprised. They developed their own confidence, even if they had to pretend.[2]

Pretending to be excited and surprised, or exaggerating feeling, had its own magic. There was a certain secrecy when it came to tourists pretending. Alcohol was again a necessary addition for this practice aboard cruises, especially as it encouraged one to laugh and scream, or cheer and snap photos at any given turn. But pretending also opened a world of play and imagination. On the cruise ships, play was most inspired by dance floors, laser tag, and pool contests. On plateau cruises, there was cellular tag, dancing-over-the-reefs nights, and craziest-picture-of-something-in-the-water contests, not to mention the traditional pool games of cruise ships past, from bikini contests to ice sculpting contests. On the beach in C-Town, there were moments when people got overly excited about a gull chasing them. It was an amazing sight: An unsuspecting group of tourists, for example, screaming and scattering, falling into the sand, or running into the water—which was arguably scarier than potentially getting pecked by a bird. But it ended in laughter, both the tourists' and Gregory's. There was a secrecy and unpredictability to it, an uncertainty to it. "Are we having fun on this boat?" It was to ask questions without having to ask them. Play was magical this way, and then out of that Gregory made his friends and won his lunch. There was something about the unpredictability in pretending or exaggerating that inspired tourists. The cruises set the tourists up, the tourists knocked themselves down, hand-feeding wildlife and eating lionfish along the way. Their play flirted with death, and they enjoyed taking advantage of it.[3]

So, when CORALREEF, which had partnered with the Wilderness Collective to help expand the coral nurseries, discovered that lionfish really loved the area too, and, in addition, that tourists were skewering them and feeding them to sharks, a new space of concern opened. Denise had her in. The lionfish were, again, a new site for her to reconnect. Denise told her colleagues that she'd been working with the community, and fishers, and tourists on lionfish issues. She offered to work as a volunteer until funds were secured for a new position. She literally had to buy time. She knew this. That was why she was the one to offer. Learning to surf the enviro-GPA-tourist world in C-Town did not entail repressing fear. It involved facing it. It was about diving into it. That was her in, in all its precarity. Denise made herself a divergence. She said she loved coral. She didn't say she really loved lionfish.

Secrets abounded.

Notes

1. "[T]o jam is to *add to*, more and more: to bring one more in, to let another in the door, add one more to the point of breaking" (Ochoa 2017: 180, emphasis original).
2. David Foster Wallace (1997) wrote critical reflections on a cruise ship, especially around the requirement to enjoy its excesses. His descriptions and points of reflection are echoed here; however, where his reflections pick up on the despairing qualities of cruise ship excesses, the story here remains open.
3. In Gregory Bateson's essay on play, he wrote: "And such adult phenomena as gambling and playing with risk have their roots in the combination of threat and play. It is evident also that not only threat but the reciprocal of threat—the behavior of the threatened individual—are a part of this complex. It is probably that not only histrionics but also spectatorship should be included within this field. It is also appropriate to mention self-pity" (1985: 134). Bateson went on to note that this metacommunicative process, which is in actuality an embodied play between ruin and possibility, is "constructed not upon the premise 'This is play' but rather around the question 'Is this play?'" (ibid.: 135).

The last time the Red Headed Woodpecker was seen in neighbouring Burlington, Ontario was when a subdivision, now called Woodpecker Woods was raised over its former habitat. (Bacher 2015)

CHAPTER 15

When I asked colleagues, locals, and interviewees about Gregory, they talked about how he lived on the Caye, but they couldn't be sure. They could only speculate. And with that speculation, stories about Gregory and the Caye, including memories, encounters, and possibilities, began to haphazardly take shape.

FTCC was tactful about taking advantage of possibility too, and there was no telling where that would go in advance. In one such instance of unpredictability, friends of the Caye entangled their friendship with a strange world. That strange world wasn't the trash pit. Rather, the strange world, in their case, was the world of science, particularly around the notion of "biodiversity."

A rise in efforts to increase biodiversity on and around the Caye had already emerged among members of the Wilderness Collective before it was taken up by FTCC. The Wilderness Collective's use of this discourse was in no way surprising. The notion of biodiversity was a powerful tool to raise awareness and support for environmental efforts, and environmental GPAs often used it. It *was* surprising, however, that FTCC members also took up the latest turn in biodiversity discourse to support their efforts. What was more, they took it up with a strategic difference.

Prior to the turn of the millennium, environmental workers in C-Town used the notion of "biodiversity" to position humans as interconnected with nature. Humans, they argued, were already part of an interconnected, natural system made up of complex relationships between diverse biological entities. If humans refused their connections to this naturally interconnected and biologically diverse system, and instead worked against them, they were deemed "others" to it. Thus, humans could be at once connected with and "othered" or threatening in relation to nature. As the years turned, the felt sense of humans' simultaneous connections and disconnections to nature increasingly generated ongoing quests for reconnection.[1]

As reconnections between humans and nature were sought, the divide between social and natural worlds blurred, and nonhumans began to slide in and out of the two categories formerly applied only to humans: either being part of the naturally interconnected system or against it. As such, not only did humans slip out of their category as other and threatening and instead become an entangled component of nature, "natural" biological entities (including lionfish or laughing gulls) could also slide into a threatening, "othered" category. And with these muddled categories, a distinction between stability and threats to it could hardly hold.

When oil extraction initiatives arrived off the coast of C-Town in the 2010s—an event that affected many neighboring countries as well—the idea of "revitalization" ahead of felt threat entered the notion of biodiversity. Biodiversity discourse no longer had any semblance of its former relationship to "good connections" versus "bad disconnections." Henceforth, the notion of biodiversity would include all biological relationships involved in revitalization, including humans, laughing gulls, and lionfish. Disconnections and their association with "otherness" were no longer threatening. Instead, all relationships were constantly calling for future connection—some more than others. This force of potential would not yet include nonliving things, like refuse or trash, however. Waste was merely waste. It got around, but it didn't call for anything. Or so it was believed.

Since they began working with the Caye, the Wilderness Collective had taken up the turn in biodiversity discourse along these lines. FTCC, then, joined in, but they did so with a difference. As was known even at that time, many FTCC members loved the Caye for its wild side, and the trash was intricately entwined with it.

With respect to the Caye, it had been discovered that active, ongoing revitalization occurred spontaneously, and it was here that FTCC members inserted themselves as friends and as other to other friends, such as the Wilderness Collective. In the process, they brought a whole new direction to the Caye's management, with their slogan "We are the Caye" and their demand to protect its natural existence. As one FTCC member wrote on their Selftory page:

"We are the Caye. The Caye is a naturally synthetic wilderness. Nature thrived by fluke, but it was no coincidence that friends thrived too."

These Caye advocates' political agenda and their attachment to the site were blatantly obvious, yet there was no sensed concern for contradiction. At one moment, they demanded to allow the Caye to emerge naturally—which confused Wilderness Collective members, among others. In another instant, they were keen to supply checklists to tour guides for plant, animal, drone, and marine life sightings, and to have the findings inventoried. Tour guides even took these checklists with them and helped tourists with identification.

Biodiversity discourse intensified active and ongoing revitalization, but for the FTCC it was a revitalization that did not just entail focusing on "biological diversity" and "nature" in a place that was so obviously unnatural. And by imagining themselves as friends of an entire, historically emergent entity, the question of revitalization moved even further from conventional notions of biodiversity than it already had. Instead, the effort to revitalize entailed access and participation in a way that demanded and imagined such participation as emerging with *what* they loved, warts and all, trash included. FTCC would still demand freedom from the Wilderness Collective's dominating disposition, even if they would have to partner with managers in the process. And for them, the freedom had everything to do with love for a toxic and trashed, yet still very much alive and proliferative caye.

Common Caye was known for contestation and FTCC got caught up in its escalation. It didn't help that the area's stimulus of pleasure and wonder simply was not powerful enough to persuade the Wilderness Collective or FTCC members to lock the Caye up, as much as the intensity was too powerful for them to actually do so. After all, the intensity was what brought the park to fruition and animals to nest. It was what brought pedestrians to wander or picnic, as well as fishers and tourists to work and hang out. It brought a tour company to make calls for redevelopment and friends to threaten to tear the place up.

It was the discovery of the red-footed booby bird nesting on Common Caye that first captured bird-lovers' attention. Given the birds' relatively large numbers on the Caye, this species, as well as other bird species, were tracked over the years. Although scientists and advocates produced new knowledge about the Caye, and documented a number of rare species, the red-footed booby statistics were what brought the attention needed to have the site recognized for its ecological importance. This species of bird, after all, was not only endangered; it was also a very desired sight by tourists. And because these birds were the most remarkable feature of Common Caye to tourists and scientists alike, they indicated that there were many more bird species, other wildlife, and landscape features to marvel over in addition, and this captured more tourist attention. At that point, the site became remarkable for its ecological significance, not only for the animals and humans living with the Caye, but also for the economy. And with that, the Caye became unique and particularly valuable as a national park.

The recognition of the site as a national park and habitat for endangered birds configured the birds as markers of environmental health, while simultaneously contributing to their natural, social, and economic value. What was more, they renewed themselves. From one moment to the next, they shifted to friends, then to resources, and then to partners. They, in other words, traveled. And with travel, their connections proliferated.[2]

Animals, like Gregory, would become representatives of this wilderness, and then markers of health, and then partners, and then resources. But that was not how I remember him, nor was it why I turned back to meet him and then returned for a visit.

I can imagine Gregory adding his fair share of mess to the friends–caye–birds–tourists mix. I can imagine him being one of those "naughty" birds who would just not budge—one of the birds so entangled in the roots and entangled piles of trash and debris that he was "locked in." The birds became so used to visitors and so stubborn that their "guests" could walk right up to them and boot them aside with a swoop of the foot, and the gulls would simply move over as if they were grudgingly conceding as much as being forced by a big shoe. He was a thoroughly knotted, troubling bird, Gregory. I knew that feeling too. But it wasn't quite right. If they were locked into anything at all, they were locked into something relentlessly moving, a feeling, "a force of lively social-material composing." And I can only imagine how Gregory felt.[3]

It didn't help that I can't be sure if Gregory and I truly met. I'm not sure if I could, if I ever did, or if it was only ever in passing. I could imagine him as being in a system understood and navigated by way of intensification. It was a navigation so utterly caught up in the marketing of "exotic" bodies to be experienced, including bodies of water, a navigation caught in capitalist flows, in the excesses of urban sprawl; and, yes, Gregory was forced to contend with them. But I cannot know how exactly he navigated his world. He was a part of it, and he was forced to represent his species, but what if he didn't move by connecting? What if he moved by soaring or gliding? I'd seen gulls ride on rising air currents. They followed the lifts and turns and sailed around and around. Where I relied on contacts, perhaps Gregory rode over them rather than with them or through them. He likely didn't get that feeling. And I could only imagine how that felt.

Yet there I was, looking to reconnect with a bird who may or may not have cared less if I did. And even then, he could rise above that connection, I imagined. I could only imagine what difference that made, if any. That where I took on a feeling, it was a matter of gliding over it for Gregory, and now I looked for him as he looked for Ian, but differently.

No one could know which currents Gregory rode to get to his caye. If he got there. I felt that we would meet again some day, even as I was not sure if he'd met me either. I didn't know what meeting was to him. I didn't know what he thought of it. If he thought anything of it at all. But I still wanted to see him again. The thing was, by remembering Gregory, by mimicking a relationship we never fully lived, we flirted with life and were taken away by it.

Although informal, FTCC's position was to let the Caye be. But good friends couldn't leave friends to just be. They desired to see them flourish, and they basked in the good company that flourishing entailed. FTCC did that too, because alongside all the requirements to connect, and all the struggles for power, they were also friends who desired to meet again. And friends can be very selfish in this regard, even more selfish than those who do the opposite. Much like parents . . . but more like friends.

Notes

1. See Celia Lowe (2006: 4) for a discussion on a turn in biological science in the US in the mid-1980s with respect to biodiversity. The reminiscence with the idea of biodiversity as it emerges in this story—it is hoped—is as curious as it is awkward.
2. See more on boundary objects in Whatmore (2002: 93). See also Chapter 5 of Whatmore.
3. The quoted phrase is from Little (2020: 133).

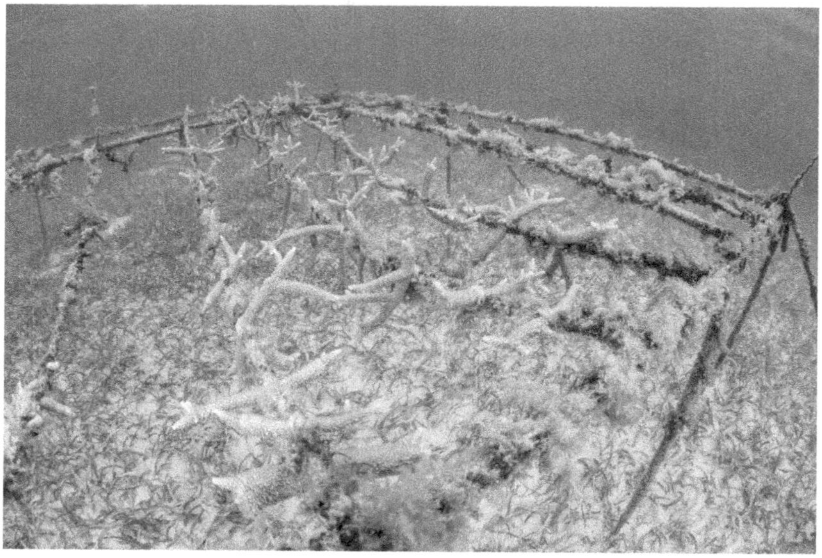

Figure 16.1. Corals grow on a metal structure for transplantation to the reef in Belize in areas where damage has occurred. © iStock.com, Velvetfish

Chapter 16

Tourism was a wandering and a wondering in C-Town. It moved into the excesses of urban sprawl and beyond its threat. It turned up in the vistas amid sprawl, appeared in expanding cities and the longing of waning townships, in hot springs run by community travelers, on board cruise ships, in the travels of retirees lined up in their RVs, in flaneur-style wanderings down city corridors. Tourists were not allies; they were not advocates, but they were not not allies or advocates, necessarily. Tourists were birds, vacationers, intrusive plant species, predatory fish habits, traveling workers, proposal writers, migrating butterflies, paper and plastics, aggregates and resources getting used up and spilling back out as garbage, memories, and sensations. How they rode that wave, how they got a feel for it, and what they were feeling, if anything at all, that was another question.

Then, as if out of the blue, Denise lost her job. All they could say was that it had to do with cutbacks. She knew, however, that with the cutbacks someone had to be let go, and they chose her because she was old, or maybe she didn't talk enough, or maybe too much. She said it hit her like a storm and the next thing she felt was being washed away into something "crazy." Had she still been working for FTCC, she would have attended events backed by her association and forged another opening. But she didn't see this one coming and so it smashed up against her, washing her away, right out of her network. She wanted back in. She should have networked when she had the chance. But Denise was tired. She wasn't up for travel. She loved her kids; she loved her home. She didn't want to leave. She'd become too focused.

Denise said that as she sat at her manager's desk getting let go, she said nothing. It wasn't until later that she felt like she was "losing it." But Denise had been through this before. She'd been hit by surprise before. She'd worked with surprises her whole life. She'd do it again. She was going with the crazy, moving into it. Just keep moving. It was kind of like accepting that so much had to get done that there was not enough time to do it, so the only option was to keep going, not in order to minimize the burn but to accept that burns happen; "shit happens," and "we'll get there when we get there, if we get there." So, keep going.

And then, with that, she said to me the last time we spoke that she was heading out to get herself some lottery tickets. Last I heard, she was working on opening her own lionfish burrito stand.

Chapter 17

I don't remember much about the Caye itself, save snippets and photos from my one visit. I had used the Caye as a stepping stone to find Gregory. I guess that it wasn't so strange that in searching for a bird I once knew I found his friends and friends who said they were friends. It was, however, rather strange that out of all the things to fight about, like a toxic trash heap, there was a problem with bird excrement on Common Caye. Much talk circulated around the matter and erupted in one key meeting, which Wilderness Collective members, FTCC affiliates, and friends attended.

Scientists and volunteers had been tracking and documenting all things Caye-related, including measuring the laughing gulls' output, that being: eggs, excrement, exploitation of resources, expanse of nesting grounds and habitat, and eruption of sound. These scientists attended the meeting and presented their findings. They called it the "Five e's of Common Caye's (In)Stability." The "five e's" was a borrowed discourse, which had emerged from Ricuesta's new e-tourism development plans. It was catchy and loaded. The scientists at the meeting had no shortage of graphs, numbers, and predictions for the future to emphasize and entice revitalization. Culling was discussed—and was rejected by appalled Wilderness Collective and FTCC members. If that was no option, something else had to happen. Nature's Last Stand would go on standing somehow.

The research that had brought about the meeting had occurred at a moment when internal divisions within FTCC were firing up. This was partly because the first FTCC members were C-Town fishers who had been fishing around Common Caye and acting as tour guides for the area. When fishers initially advocated for the Caye, some saw their emergence as a group as the beginnings of revitalizing C-Town's fishers' cooperative. When the Wilderness Collective took over, not only did many FTCC members view the Collective's actions as stealing their beloved caye out from under them, they saw the partnership as a domination that once again disregarded fishers' identity, their lifeways, and their views of the Caye, including their relationships with it and each other. This was especially the case because they were increasingly losing access since the Caye had become a national park. The loss of access prompted tensions between families—tensions that often erupted as attacks and claims of irresolvable differences—to flare yet again. When it came to FTCC relations, though, these escalating struggles were not expressed as a source of internal tensions. Rather, the eruption occurred in tensions between FTCC and the Wilderness Collective as they struggled over the Caye's management and the desire to let it be.

To make matters more complicated, the issue of gull excrement had also become a national issue, through which Nature's Last Stand reappeared. Scientists argued that Common Caye proved that nature perseveres and requires respect, and that all members needed to mind the "five e's of e-thnicity." The five e's came up repeatedly because environmental efforts and sustainability through the promotion of e-tourism were increasingly becoming key to economic sustainability in Ricuesta, while it maintained the image of an independent, sustainable, and unified state. The goal, which was covered over with discourse on tourism development, was simply code for "sustainability," except that in Ricuesta, "sustainability" was not geared toward some utopian dreamworld. Instead, it was geared toward weathering the storm, persevering despite adversity. And perseverance in such a case was to keep going. Nature was caught up in that same discourse, and partnerships geared toward revitalization had become a way to persevere through the very precarious conditions of life in C-Town, economic crisis, and environmental degradation. The environment was a major site through which this feeling was driven further and reentangled with the excesses of urban sprawl and tourism.

"E-thnicity" discourse had already emerged with the rise of e-tourism and was a way to refer to and tie the many Ricuestian communities, with their various histories, to the national aim of moving toward a fully touristic "e-state." The five e's, terms that showed up in almost every document devoted to tourism development, were as follows: to promote **E**quality, to be fully **E**lectronic, to develop renewable and self-sufficient sources of **E**nergy, to move toward **E**nvironmental sustainability, and to promote **E**-tourism. Each "e-" was more of a method of weathering the storm than a vision for the future. The call for equality and the call for environmental sustainability aligned with the making of public–private partnerships and cross-industry connections. This first "e" in particular became a way to promote unity despite differences in identity—the claim being, we are different but all potentially partners. This, of course, covered over unequal power relations within partnerships, but the idea stuck nonetheless. The move toward self-sufficient sources of energy and the decades-long struggle to be fully electronic were aimed at diving into and supporting travel and movement, including that of finance, bodies, goods, and resources, and the promotion of e-tourism drew the previous four together through tourism development. The Caye entered directly into this discourse. It had emerged not only because it had persevered through neglect and degradation, through storms and ongoing struggle; it also persevered because it emerged alongside the rise of cruise ship tourism. It thus stood as a symbol of perseverance and sustainability not only for nature but for the nation.

This was not the first time that environmental concerns had emerged alongside the emergent tourism sector and nationalist concerns. In fact, tourism and environmental interests had overlapped since prior to when the struggle for independence began. The first "tourists" to Ricuesta started visiting in the 1930s, mostly to hunt exotic animals or fish. This activity had occurred before the 1930s as well, but it was during the Great Depression that something of a market emerged, which specifically drew wealthy men to hunt and stay in lodges built for them.

At this time, concern around the hunting, especially of exotic animals, had already been voiced, but it was largely suppressed by colonial authority. As the possibility of independence gained momentum through the 1960s, local environmental organizations started to appear, and their efforts began to be acknowledged. At the same time, tourism emerged once again in the form of a small—albeit growing—group of visitors interested in experiencing "the wild." In the ensuing years, the Ricuestian government began to partner with environmental organizations to help fill gaps where resources or time for matters concerning the environment were limited. The Ricuestian Conservation Assembly (RCA) became one such partner and the first organization to support a combination of environmental protection and tourism.

What made the RCA very interesting and different was the fact that, as a nongovernmental organization, this group of nature enthusiasts did not simply emerge in a supposedly unknown region of the globe. Rather, they were thoroughly connected with a vast and expanding network of conservationists, which was global in scope. The RCA, since its beginnings, was made up of a network of connections between environmental advocates from all over the world, locals, political leaders, civil servants, international donors, other nongovernmental organizations, advocates, tourists, volunteers, and more. Even with little infrastructure for easy networking, this group spread images and their messages transnationally, contacted foreigners for information and support, secured sources of international funding, established worldwide alliances, and used contacts for help with lobbying efforts.

To the world beyond Ricuesta, their focus was on conserving natural resources for the sake of the planet. Internally, their quest for conservation was also caught up in a struggle for independence from colonial control—control that had relied on resource removal through force and oppression. Aligned with the goal of conservation, the RCA began recommending new practices to help diversify the economy by promoting conservation and wildlife growth over resource removal. They promoted the development of ecotourism—an industry that would work to support a newly emergent nation threatened by the risk of falling into disrepair as it entered the global arena and its crushing pressures.

Environmental efforts and sustainability through the promotion of ecotourism increasingly became key to economic growth through the ensuing decades. There were many goals, much deliberation, and many shifts in alliances, negotiations, turns in funding, and power struggles along the way for all involved with conservationist endeavors. The partnerships moved through very precarious conditions of life in the emergent nation, especially in the event of global economic downturns. Tourism grew exponentially over the years, and as more tourists showed up, the environment faced further degradation. With more degradation, more tourists showed up, people moved further and further in search of work as costs of living increased, tensions between families and between communities grew, and natural environments struggled even more.

Still, environmentalists tried to work with these constraints as they fought tirelessly to align ecotourism with sustainability. They built real relationships with each other, their partners, politicians, and wildlife through crisis after crisis. They had to. And as tourism became increasingly dominant in Ricuesta, the partnerships and relationships made continued to multiply and shift alongside precarious natural, social, political, and economic environments, as everyone held out hope that ecotourism would help lessen the heightening sense of ongoing environmental and economic crisis.

During this time of heightening crisis, even nationalists hoped that tourism could help the nation's progress by controlling the tensions experienced between Ricuesta's many communities. They anticipated that the tourism sector could help align the points of division between groups and communities—their various lifeways and differences—with a common and promising goal for Ricuesta: to showcase the nation's diverse ties to the land and landscapes in a common celebration for tourists to see. It was hoped, in other words, that tourism would control tensions and disconnects that crisis after crisis so proficiently intensified.

It could not.

Yet diversity was always key to networking, and it was something many environmental groups worked with as members continuously opened their organizations to connect with others over the years. Distinctiveness and disconnection, in these early years, were already a source of possibility rather than threat. This was because, from the start and counter to earlier hopes for tourism, environmental organizations of Ricuesta did not try to contain difference. Disconnection was never the goal, but it was not the end of the world either. They accepted it. Enviro-NGOs, with all their knotted and contradictory connections, had figured out a way to weather the storm of unpredictability. And that was celebrated as much as it emerged to be troublesome, to be different. As troublesome as it was, it was certainly encouraging when differences emerged with diverse and new ways to support each other, the environment, and the nation.

By the 2020s, the e- of e-thnicity and of e-state discourse took on the coincidental similarity of a letter and linked ethnicity back to the e-tourism industry again. There emerged a "touristic e-thnicity" where tourism and ethnicity were brought together again in a new state discourse. Hence, the Wilderness Collective called attention to perseverance and e-thnicity. And by doing so they were calling out a need to cooperate and focus on ways to persevere, and that perseverance, they knew, was through environmental revitalization. In promoting the Caye's revitalization, they called on e-state discourse to bring differences in identity and desire, tourism and the environment into new partnerships and opportunities for "all Ricuestians."[1]

It wasn't quite the contradiction, then, that so much trouble emerged because the gulls kept persevering too. FTCC had set up decoys to scare off the birds, but the birds kept reproducing. They offered new areas to nest, but the birds kept returning and reproducing. They even tried removing eggs from their nests, but they kept reproducing. Gregory persevered through it too, I imagine.

In response to the gulls' stubborn behavior, divisions appeared across multiple parties. On one hand, it was argued that the laughing gulls were ruining nature, while others argued management was. In the end, the Wilderness Collective argued that much could be learned about nature in this circumstance, but it didn't mean toxicity should be ignored and waste left to its own devices. Some form of intervention was the best way forward, that was, if Nature's Last Stand was to continue standing with "all Ricuestians," they argued. And they were the best fit for the position. Of course, the idea sparked debate within FTCC itself, and the group reemerged as two groups again, one formal and the other informal, with the FLBCC separating entirely. Surprisingly, some Wilderness Collective members joined the informal group; some formerly informal members joined the formal one; and some joined both. It was complicating to be in a complicated relationship with a toxic heap of trash and trees and other life-forms thriving with it, building onto it. It complicated everything to fall for the twisted nature of the Caye, and to invite another return.

Once the informal FTCC and its alliance with FLBCC reemerged, an old argument came forward to rear its ugly head again. Without FTCC's official status as a GPA, they'd be beyond oversight. It especially didn't help that FTCC's informal side seemingly acted on impulse at times. How could the Wilderness Collective, their partners, and other organizations and associations along the coast coordinate their intentions and interconnected activities for environmental revitalization? How could they manage under these conditions? How could they enforce the rules and regulations required for the revitalization of the reefs, and now other bird species on the Caye? At the time of my visit to the Caye, there were ongoing debates on the laughing gull issue.

I imagine interviewing Gregory during these happenings. I imagine he had a different way of dealing with the loss of his clutch—that he approached it not through fear and sadness but anger and exasperation.

"Bastard tools ate my eggs!" he yells in an interview. "So, we had to lay another that season. Do you have any idea how difficult that is? Just out of spite, we laid six. And so late in the season too. Who hunts at the end of the season? I mean, come on! I am going to shit all over everything; they are so going to regret washing their hair ever—"

"Gregory," I interrupt. "I don't think they ate—"

"They ate them!"

"They're talking about cleaning up the waste to neutralize the—" I say before getting cut off again.

"They want to talk about waste?" Gregory cries. "Do they really want to talk about toxic waste?!"

Gregory raises his voice louder and louder until it is a full-blown laughing squawk. He isn't laughing, but the sound cuts through as laughter until I can hear nothing but. His voice overpowers my own thoughts to the point that I can't even say a word. All I can do is watch. And then, just as it can't get any louder, he starts talking in circles while turning around and around on his straight, skinny legs and webbed little feet. He guffaws, "They want to talk to me do they want to? Any idea how hard? At the end of season. They want to talk to me; unbelievable, they want to—"

"But Gregory," I finally interject; "You were offered another caye!"

"Five miles further away?!" And the squawking takes off again.

I can't help but burst out laughing, but it isn't laughter.

No one knew in advance exactly where the pull of friendship would fall. Bonds formed between one and another, and at the expense of yet another standing right there. "We need to consider the mangroves," one observer said. "There is only so much toxicity they can handle. Don't they count? Don't the herons count? How about coral?"

The gulls likely did not care much about the state of the mangroves or anyone's view of the place. Different life-forms tended to have other ideas in mind. They refused to stay still for long enough to be counted and forced their census to reach more than what was possible on a tiny island; they outsmarted the endeavors of their featherless, two-legged associates; they even grew callous and left their friends practically crying, "But why won't you let me help you?!"

The Caye was made in excess, and it lived by it every step of the way. Stories of the site were no different. They remained rich with humor, silliness, and puns that put a spin on things. They were also thick with touching descriptions, passionate displays of longing, and explosive confrontations. They entangled with histories of valuing certain groups over others, as much as they comprised moments in which one friend was deemed the worst in the world, whereas those doing the prodding, distressing, culling, and booting aside with a big shoe were not. There were those who took up arms in the debates; and there were those seemingly unaware that they had just let themselves in, crashed on their friend's couch, and believed it would be "as if they weren't even there" if they tiptoed around and "left only footprints."

It seemed there was always an excess to the friendships, excesses that ruined it for everybody and forced them to "begin" again in the middle of things.

Bonds between the Caye, its embodied others, and friends, thus, were stretched thin. The tensions the gulls produced provoked all sorts of creative and bizarre ideas, from poking at them, to setting up inflatable, flailing-arm Halloween decorations that would also laugh maniacally at any given movement, to unleashing drones to scare them off.

The laughing gulls were laughing as friends had a way of laughing at each other in C-Town. Friends had their secrets; they spoke and acted beyond their loyalties; they used each other as a crutch, dreamed through each other too, lost grip on reality, and then were sorely disappointed by reality. They had a funny way of slipping away, then reappearing out of the virtual blue with a "happy birthday" comment. Friends listened, but they did not fully hear. They agreed with nodding smiles, with a loving disregard. They involved exercises in futility.

They included moments where girl met bird and then whacky stories got written about a friend.

Notes

1. This section circles in very odd ways around the notion of "touristic ethnicity," which Teresa Holmes (2010) has discussed in the context of Belize. Using a concept introduced by R. E. Wood, Holmes builds on the idea of tourist ethnicity by linking its emergence to historical concerns surrounding ethnicity and the state in Belize. This section picks up on her analysis of this notion, the threat of divisiveness, and the idea of the tourism industry as a "safe zone" (following Wilk 1995: 128, cited in Holmes 2010: 157). For comparison, see Holmes (2010: especially 157–58).

Chapter 18

On a humid day in September, I visited the Caye to see where Gregory might have lived. Despite decoys, despite the use of small drones set to randomly buzz at or drop balls on them, despite volunteers yelling and running at them with their arms flailing, the gulls continued to go about their daily routines. That day I visited, there were more gulls than ever.

When it was time to leave, our tour guide led our group down the tourist path, away from other wildlife and back toward his e-boat. As we walked up the makeshift beach, a laughing gull crossed the tour guide's path and blocked our way. It stood there looking, its head turning side to side. The tour guide stopped in his tracks and paused for a moment, looking at the gull. Then, he pushed at the animal standing in his way with the side of his shoe. The bird merely leaned to the side and then, with its wings slightly lifted and head jutting forward, scurried in a circle, up over the shoe and back around to where it was prior to the push. Everyone stopped and watched. Then, as if nothing was out of the ordinary, the guide carried on, walking around the little sightseer. The tourists followed.

In the face of sure claims of betrayal, friends had a way of troubling what it was to love. When friends became embodiments of a dream, they had a funny way of demanding respect. This was because friends did not fulfill dreams; they were shoulders to cry over.

Memories of Gregory emerged as an event of telling in which relations of friendship were made and unmade in the midst of tourist sprawl and getting chased down the beach. I was taken away by memories, and in the process, they conjured a return. Friendships were adventures, rites of passage even—if a rite of passage was a dramatic performance consisting of a state in which play was the name of the game, and through which friends met again, but differently.

The memory of the chase took me back and then released me into a fall. It opened a powerful moment that made something in the memorable agitations it reignited, that came back unexpectedly, and differently, as in the story of the water's-edge friendships. And this also provided the chance to think about a question still lingering. It was a wonder that if inhumanity is grounded in alienation not only from ourselves but also from each other, why was it that in the potentializing forces inherent to the very act of working with and working on relations as they are felt and reorganized in experience, and with each other—why was it specifically there, in the excesses of hierarchical arrangements, in immanent relating, in the very act of being so sincerely creative, that so much difficulty directly arose?

If only relationships were easy.

Some might've said that friendships and the stories they shaped were tools for becoming other or becoming different. That was what I thought might be the case, except perhaps it was also useful to think of them as events of letting go.[1]

What happened to friendships emerged across the places through which travel occurred, and if they were remembered. A beginning was not really the origin but a source of movement and difference. For me, friendships emerged in memories recalled of those who got away, who I had to let go, of Gregory, who complicated my sweet memories of a beach. And it had to be done in whatever ways possible in this world, even if it meant bringing together pieces of multiple encounters, stories, histories, pillaged, pulled, folded in, and dug out of every experience that memories could conjure. As for Gregory in these travels: A bunch of tourists wandered to the water's edge not knowing what to expect. Memories of those who had also stumbled upon the place followed. At that moment, previous priorities and attachments did not disappear but expanded, that is, expressly intensified across multiple bodies and worlds, and conventions of who or what made a friend were ruined; standing there, in the shaky calm of an oncoming storm, we began again . . .

I remember standing on the beach picking up on the smell of vanilla caught up in the wind as a storm fast approached C-Town. From there, memories, and memories of memories of Gregory and Common Caye, came back around. I wondered what we would come back to and what would come back to us once the storm came through. It was a wondering about what the beach would be like, and what we would remember then.

Figure 18.1. Charlie the night heron, Toronto waterfront. © Ines Taccone

Note

1. Interestingly, in Hannah Arendt's (1998) *The Human Condition*, there is something that emerges in the space of appearance, which Arendt referred to as a "regard" from a distance—specifically "the distance which the space of the world puts between us" (ibid.: 243). With no predetermined or formal connection, friendship enacts such a regard—it attends to difference over predetermined connections with an intention to watch and listen despite this lack, not because of lack. This regard can be described as an affective moment of potential connection—a moment in which a connection across space and because of difference might be made—yet with an interesting difference. The difference, following Arendt, is in the desire to see the other act. Rather than a sensation of potential connection, then, the desire is in seeing the other become more than they are, let go, or even do something new. It is to accept disconnection as a form of collaboration without hinging on connection. See also Arendt (ibid.: Chapter 5).

References

Arendt, Hannah. 1998. *The Human Condition*, with an Introduction by Margaret Canovan, 2nd edn. Chicago: University of Chicago Press.

Bacher, John. 2015. "Urban Sprawl Keeps Killing Off What is Left of Niagara's Wildlife—Get Engaged in Ontario's Greenbelt Review." *Niagara at Large*, 21 March. Retrieved 8 November 2022 from https://niagaraatlarge.com/2015/03/21/urban-sprawl-keeps-killing-off-what-is-left-of-niagaras-wildlife-get-engaged-in-ontarios-greenbelt-review/.

Barad, Karen. 2003. "Posthumanist Performativity: Toward an Understanding of How Matter Comes to Matter." *Signs: Journal of Women in Culture and Society* 28(3): 801–31.

———. 2007. *Meeting the Universe Halfway: Quantum Physics and the Entanglement of Matter and Meaning*. Durham, NC: Duke University Press.

Bateson, Gregory. 1985. "A Theory of Play and Fantasy." In *Semiotics: An Introductory Anthology*, ed. Robert E. Innis, 129–44. Bloomington: Indiana University Press.

Belsky, Jill M. 1999. "Misrepresenting Communities: The Politics of Community-Based Rural Ecotourism in Gales Point Manatee, Belize." *Rural Sociology* 64(4): 641–66.

Benjamin, Walter. 1968. *Illuminations: Essays and Reflections*, ed. Hannah Arendt, trans. Harry Zohn. New York: Schocken.

Bergson, Henri. (1913) 2001. *Time and Free Will: An Essay on the Immediate Data of Consciousness*, trans. F. L. Pogson. Mineola, NY: Dover.

Berlant, Lauren. 2011. *Cruel Optimism*. Durham, NC: Duke University Press.

Biehl, João, and Peter Locke, eds. 2017. *Unfinished: The Anthropology of Becoming*. Durham, NC: Duke University Press.

Brondo, Keri Vacanti. 2013. *Land Grab: Green Neoliberalism, Gender, and Garifuna Resistance in Honduras*. Tucson: University of Arizona Press.

Buck-Morss, Susan. 1991. *The Dialectics of Seeing: Walter Benjamin and the Arcades Project*. Cambridge, MA: MIT Press.

———. 1992. "Aesthetics and Anaesthetics: Walter Benjamin's Artwork Essay Reconsidered." *October* 62: 3–41.

Busiello, Antonio. 2022. "A Coral Disease is Spreading through World's Second Largest Barrier Reef." *SCUBA Diving*, 14 June. Retrieved 8 November 2022 from https://www.scubadiving.com/coral-disease-is-spreading-through-worlds-second-largest-barrier-reef.

Cvetkovich, Ann. 2003. *An Archive of Feelings: Trauma, Sexuality, and Lesbian Public Cultures*. Durham, NC: Duke University Press.

Deleuze, Gilles. (1962) 1993. *Difference and Repetition*, trans. Paul Patton. New York: Columbia University Press.

———. 2006. *Nietzsche and Philosophy*, trans. Hugh Tomlinson. New York: Colombia University Press.

Deleuze, Gilles, and Félix Guattari. 1987. *A Thousand Plateaus: Capitalism and Schizophrenia*, trans. Brian Massumi. Minneapolis: University of Minnesota Press.

Duffy, Rosaleen. 2005. "The Politics of Global Environmental Governance: The Powers and Limitations of Transfrontier Conservation Areas in Central America." *Review of International Studies* 31(2): 307–23.

Fairhead, James, and Melissa Leach. 2003. *Science, Society and Power: Environmental Knowledge and Policy in West Africa and the Caribbean*. Cambridge: Cambridge University Press.

Gordon, Avery. 1997. *Ghostly Matters: Haunting and the Sociological Imagination*. Minneapolis: University of Minnesota Press.

Grandia, Liza. 2007. "Between Bolivar and Bureaucracy: The Mesoamerican Biological Corridor." *Conservation and Society* 5(4): 478–503.

Gregg, Melissa, and Gregory J. Seigworth. 2010. "An Inventory of Shimmers." In *The Affect Theory Reader*, ed. Melissa Gregg and Gregory J. Seigworth, 1–25. Durham, NC: Duke University Press.

Haraway, Donna J. 2008. *When Species Meet*. Minneapolis: University of Minnesota Press.

———. 2016. *Staying with the Trouble: Making Kin in the Chthulucene*. Durham, NC: Duke University Press.

Hayward, Eva. 2010. "FingeryEyes: Impressions of Cup Corals." *Cultural Anthropology* 24(4): 577–99.

Holmes, Teresa. 2010. "Tourism and the Making of Ethnic Citizenship in Belize." In *Tourism, Power, and Culture: Anthropological Insights*, ed. Donald V. L. Macleod and James G. Carrier, 153–73. Bristol: Channel View.

Hustak, Carla, and Natasha Myers. 2012. "Involutionary Momentum: Affective Ecologies and the Sciences of Plant/Insect Encounters." *Differences: A Journal of Feminist Cultural Studies* 23(3): 74–118.

Igoe, Jim, and Dan Brockington. 2007. "Neoliberal Conservation: A Brief Introduction." *Conservation and Society* 5(4): 432–49.

Ivy, Marilyn. 1995. *Discourses of the Vanishing: Modernity, Phantasm, Japan*. Chicago: University of Chicago Press.

Little, Kenneth. 1991. "On Safari: The Visual Politics of a Tour Representation." In *The Varieties of Sensory Experience: A Sourcebook in the Anthropology of the Senses*, ed. D. Howes, 149–63. Toronto: University of Toronto Press.

———. 2020. *On the Nervous Edge of an Impossible Paradise: Affect, Tourism, Belize*. New York: Berghahn. Retrieved 7 February 2023 from Proquest Ebook Central.

Lowe, Celia. 2006. *Wild Profusion: Biodiversity Conservation in an Indonesian Archipelago*. Princeton: Princeton University Press.

Manning, Erin. 2016. *The Minor Gesture*. Durham, NC: Duke University Press.

Massumi, Brian. 2002. *Parables for the Virtual: Movement, Affect, Sensation*. Durham, NC: Duke University Press.

McLean, Stuart. 2011. "Black Goo: Forceful Encounters with Matter in Europe's Muddy Margins." *Cultural Anthropology* 26(4): 589–619.

———. 2017. *Fictionalizing Anthropology: Encounters and Fabulations at the Edges of the Human*. Minneapolis: University of Minnesota Press.

Medina, Laurie Kroshus. 1999. "History, Culture, and Place-Making: 'Native' Status and Maya Identity in Belize." *Journal of Latin American Anthropology* 4(1): 133–65.

——— . 2003. "Commoditizing Culture: Tourism and Maya Identity." *Annals of Tourism Research* 30(2): 353–68.

———. 2010. "When Government Targets 'the State': Transnational NGO Government and the State in Belize." *PoLAR: Political and Legal Anthropology Review* 33(2): 245–63.

———. 2015. "Governing through the Market: Neoliberal Environmental Government in Belize." *American Anthropologist* 117(2): 272–84.

Ochoa, Todd Ramón. 2007. "Versions of the Dead: *Kalunga*, Cuban-Kongo Materiality, and Ethnography." *Cultural Anthropology* 22(4): 473–500.

———. 2010a. "PRENDAS-NGANGAS-ENQUISOS: Turbulence and the Influence of the Dead in Cuban-Kongo Material Culture." *Cultural Anthropology* 25(3): 387–420.

———. 2010b. *Society of the Dead: Quita Manaquita and Palo Praise in Cuba*. Berkeley: University of California Press.

———. 2017. "Origami Conjecture for a *Bembé*." In *Crumpled Paper Boat: Experiments in Ethnographic Writing*, ed. Anand Pandian and Stuart McLean, 172–84. Durham, NC: Duke University Press.

Pandian, Anand. 2019. *A Possible Anthropology: Methods for Uneasy Times*. Durham, NC: Duke University Press. Rakuten Kobo edn.

Pandian, Anand, and Stuart McLean. 2017. "Prologue." In *Crumpled Paper Boat: Experiments in Ethnographic Writing*, 1–10. Durham, NC: Duke University Press.

Pine, Jason. 2019. *The Alchemy of Meth: A Decomposition*. Minneapolis: University of Minnesota Press. Rakuten Kobo edn.

Piot, Charles. 2010. *Nostalgia for the Future: West Africa after the Cold War*. Chicago: University of Chicago Press.

Raffles, Hugh. 2010. *Insectopedia*. New York: Pantheon.

Roessingh, Carel, and Karin Bras. 2003. "Garifuna Settlement Day: Tourism Attraction, National Celebration Day, or Manifestation of Ethnic Identity?" *Tourism, Culture & Communication* 4(3): 163–72.

Stewart, Kathleen. 1996. *A Space on the Side of the Road: Cultural Poetics in an "Other" America*. Princeton: Princeton University Press.

———. 2007. *Ordinary Affects*. Durham, NC: Duke University Press.

———. 2008. "Weak Theory in an Unfinished World." *Journal of Folklore Research* 45(1): 71–82.

———. 2010. "Worlding Refrains." In *The Affect Theory Reader*, ed. Melissa Gregg and Gregory J. Seigworth, 339–54. Durham, NC: Duke University Press.

Stinson, James. 2014. "Mother Nature's Best Kept Secret? Exploring the Discursive Terrain and Lived Experience of the Ecotourism-Extraction Nexus in Southern Belize." In *The Ecotourism-Extraction Nexus: Political Economies and Rural Realities of (Un)Comfortable Bedfellows*, ed. Bram Büscher and Veronica Davidov, 88–109. New York: Routledge. Retrieved 7 February 2023 from Proquest Ebook Central.

———. 2017. "The Will to Conserve? Environmentality, Translation, and Politics of Conservation in Southern Belize." Ph.D. dissertation. Toronto: University of Toronto. Retrieved 16 January 2023 from https://tspace.library.utoronto.ca/bitstream/1807/79521/3/Stinson_James_201706_PhD_thesis.pdf.

Sutherland, Anne. 1998. *The Making of Belize: Globalization in the Margins*. Westport, CT: Bergin and Garvey.

Taussig, Michael. 1993. *Mimesis and Alterity: A Particular History of the Senses*. New York: Routledge.

Tsing, Anna Lowenhaupt. 2005. *Friction: An Ethnography of Global Connection*. Princeton: Princeton University Press.

———. 2015. *The Mushroom at the End of the World: On the Possibility of Life in Capitalist Ruins*. Princeton: Princeton University Press.

Wallace, David Foster. 1997. *A Supposedly Fun Thing I'll Never Do Again: Essays and Arguments*. New York: Little, Brown.

Whatmore, Sarah. 2002. *Hybrid Geographies: Natures, Cultures, Spaces*. London: Sage.

Yusoff, Kathryn. 2013. "Insensible Worlds: Postrelational Ethics, Indeterminacy and the (K)nots of Relating." *Environment and Planning D: Society and Space* 31: 208–26.

Index

Printed in the USA
CPSIA information can be obtained
at www.ICGtesting.com
JSHW052028050823
45513JS00007B/15

9 781805 390343